CULTURE IN CONFLICT

CULTURE IN CONFLICT

*Irregular Warfare, Culture Policy,
and the Marine Corps*

Paula Holmes-Eber

Stanford Security Studies
An Imprint of Stanford University Press
Stanford, California

Stanford University Press
Stanford, California

Printed in the United States of America on acid-free, archival-quality paper

Library of Congress Cataloging-in-Publication Data

Holmes-Eber, Paula, author.
    Culture in conflict : irregular warfare, culture policy, and the Marine Corps / Paula Holmes-Eber.
    pages cm — (Stanford security studies)
    Includes bibliographical references.
    ISBN 978-0-8047-8950-9 (cloth : alk. paper)
    ISBN 978-0-8047-9189-2 (pbk. : alk. paper)
    1. United States. Marine Corps. 2. Cultural awareness—Government policy—United States. 3. Intercultural communication—Government policy—United States. 4. Irregular warfare—United States. 5. Organizational change—United States. 6. Sociology, Military—United States. I. Title.
VE23.H57 2014
359.9'64—dc23

2013044154

ISBN 978-0-8047-9190-8 (electronic)

To my mentors—
Helen Schwartzman
and
Jeffery Bearor
who probably had no idea how seriously
I would take their advice to "study up"

Contents

Figures and Tables

Figures

Tables

Acknowledgments

ALL BOOKS ARE ULTIMATELY THE RESULT OF MANY people's assistance. However, this book, more than any other project I have worked on, has truly exemplified the Marine Corps ethos of "there's no 'I' in team." So many people, in fact, have supported this research, participated in the study, and given commentary on my drafts that I feel almost embarrassed to claim this work as my own. Although there is not space here to give credit to all the amazing people who have helped me see the world through Marine Corps eyes, there are several who do need special mention:

Colonel (ret) Jeffery Bearor (SES), who had the vision to support this project from its inception, provide extensive mentorship throughout the seven years of my research, and faithfully read and comment on many drafts. This book truly is the result of his tireless efforts on my behalf.

Lieutenant General (ret) Paul van Riper, who—from the very first day we met—challenged and pushed me to think critically and deeply about the way that the Marine Corps was responding to its cultural challenges. For his extensive and detailed reading and commentary on the entire book (peppered by an amazing knowledge of Marine Corps history and lore) I am truly grateful.

Colonel (ret) Jerre Wilson, who understood that research, writing, and teaching are inseparable, and dedicated the time, resources, and support necessary to help me complete this project, all as part of my duties as professor of operational culture at Marine Corps University.

Colonel (ret) George Dallas, from whom I probably learned most about Marine identity, leadership, and the challenges of organizational change. Through his personal example and our many, many conversations I have had the rare opportunity to truly look at the world from a Marine's perspective.

The staff at the Marine Corps Center for Advanced Operational Culture Learning (CAOCL) and especially Gunnery Sergeant (ret) Alex Mesa, Hamid Lellou, and Rashid Qawasmi for taking time out of their busy schedules to let me observe their training classes and discuss the challenges of teaching culture and language in the Marine Corps.

Colonel Mark Desens, Colonel (ret) Royal Mortenson, Colonel Kris Still-ings, Colonel Alex Vohr, Colonel Tracy King, and Colonel Christopher Wood-bridge. I am deeply indebted to the directors of the Marine Corps Officer Candidates School (OCS), The Basic School (TBS), Expeditionary Warfare School (EWS), Command and Staff College (CSC), School of Advanced Warf-ighting (SAW), and Marine Corps War College (MCWAR) for allowing me to observe classrooms and exercises and interview faculty and students for this study. Furthermore, several of these directors took time out of their busy schedules not only to discuss their schools and their programs but also to provide critical commentary on draft chapters of this book.

Colonel Rickey Grabowski, Colonel Benjamin Blankenship, Colonel Rob-ert W. Jones, Colonel Daniel Wilson, Sergeant Major Gary Buck, Gunnery Sergeant Shawn Potvin, and Captain Will Patrone for allowing me to tromp all over the Marine Corps Recruit Depot at Parris Island, North Carolina, and showing me how the Marine Corps makes a Marine.

Kerry Fosher, my anthropological partner in crime, with whom I have shared countless hours working on the Marine Corps culture venture and whose own rich perspectives have greatly informed my own thinking on the subject.

Allison Greene-Sands, who always was willing to provide her higher-level perspective about the problems of organizational change and the challenges of policy implementation from the point of view of the Department of Defense community.

Andrea Hamlen and Stase Rodebaugh, who tirelessly edited every page of this book while working frantically during the peak rush of master student thesis submissions.

Geoffrey Burn, James Holt, and Tim Roberts, my editors at Stanford University Press, who demonstrated truly extraordinary patience, persistence, and professionalism in working with me to prepare this book for publication.

Deborah Wheeler, dearest friend, scholarly colleague, partner in juggling roles as professor and mother, and tireless cheerleader, who read every unpolished initial page and kept me going with her unbeatable optimism and humor.

Disclaimer

To protect the identity of all participants in this study, all personal names are pseudonyms. Where rank and billet (position) could identify the speaker, Marines' titles and ranks may also be altered, along with gender, age, or any other identifying demographic information. The reader should therefore be extremely cautious in drawing any ties between the speakers and actors in this study and any individuals, alive or dead, who may resemble the Marines in this study.

The opinions expressed in this book are the author's own and do not represent those of Marine Corps University, the Marine Corps Center for Advanced Operational Culture Learning (CAOCL), Training and Education Command (TECOM), the U.S. Marine Corps, the Department of Defense, or the U.S. government. All omissions, errors, and misinterpretations are completely the responsibility of the author.

Prologue

L ATE ON A DRY DUSTY OCTOBER AFTERNOON, THE
Marines of 4th CAG (civil affairs group)[1] came to a slight rise on
the Somali plains. For the past six months the company had been conducting civil affairs and humanitarian aid operations across the Horn of Africa
(HOA)—building schools, clinics, wells, and roads and inoculating ani-
mals—in order to foster economic and political stability in the region. Their
task had been anything but simple: as part of a HOA Joint Task Force, a total
of forty-five civil affairs Marines, along with a handful of engineers, medics, and veterinarians, were expected to cover an area almost two-thirds the
size of the continental United States in a region of hundreds of different languages, cultures, and warring ethnic groups.

On the other side of the rise they could see the village of Mahmadiyya.
The agriculturally based settlement had been suffering a drought for the past
five years. In writing his after action report about the operation, the task force
commander Colonel Franklin (pseudonym) described the village's poverty-
stricken situation, "The only source of water [was] a stinkin' muddy river,
full of crocodiles and filth, about two miles from the people." The unit's first
response was, as the commander noted, "to get the well drillers in there, go
down about 600' and provide free water to all who wanted it."

Colonel Franklin quickly realized, however, that this was a "bad move.
About 10 percent of the population makes their meager living hauling water

from the muddy river." Building a well, he observed, would only result in "instant unemployment and resentment. So what to do?"

I look at the U.S. Marine, Army, Navy, Air Force, and international military officers in my Operational Culture class at Marine Corps University. We are using a real case study to think through the challenges of applying cultural understanding to improve military operations. "What would you do?" I ask, challenging the students to think through the problem from the colonel's perspective.

The students hesitate for a moment. So I lead the discussion by reading the conclusion of Colonel Franklin's after action report describing the Marines' operations in Mahmadiyya: "Restraint. We let the people keep using the bad water, even though we knew we could do better, because the cure will be worse than the disease."

"Was this the only solution?" I toss the question back to the majors and lieutenant colonels in the seminar. Given the extensive deployment experience of most of these master's degree candidates, I anticipate there will be some lively debate here. Some think the colonel made the only reasonable decision. If the purpose of the civil affairs group was to foster greater stability in the region, then increasing unemployment certainly was not a good move.

"Maybe the unit could have hired the water carriers to build the well," suggested a Marine lieutenant colonel. "That would only be a temporary fix," noted his Navy colleague across the table. "Perhaps the unit could have come up with a jobs program for them," added another student. "But we're not in the job creation business," argued the major across from him. "That's what AID [U.S Agency for International Development] and the NGOs [non-governmental organizations] are there for."

"Sure," an Army ranger noted dryly. "How many NGOs do you see running around the field when the bullets are flying? We're the ones stuck with the job, whether we like it or not."

There is silence in the room as the officers all think back to the frustrating, seemingly insolvable cultural and human problems they faced in their last deployments—the daily expanding expectations of the U.S. public and government that the Marine Corps and its sister services were responsible not only for conducting wars but also for solving all the problems of broken countries, from building wells to creating stable governments.

Suddenly, a hand shoots up. "Ma'am, maybe they could build the well, but put it a mile or so from the village, closer to the river." The officers in the classroom look at him, not quite convinced. "That way the people in the village could get clean water," continued the student, a Marine major and engineer who had recently returned from deployment to Afghanistan. "They wouldn't get eaten by crocodiles. And the water carriers would still have a job."

The officers are intrigued. The solution seems inefficient to time-obsessed American eyes. And yet despite its unconventional and indirect method, from the point of view of the Somali villagers, the solution would in fact give them water without disrupting their economy.

This book is about cross-cultural problem solving—about the messy process of translation, interpretation, and program implementation as two different worlds struggle to make sense of each other. The focus is not upon the answer, but the process. *Culture in Conflict* focuses on what happens to a policy or program, created in Washington, D.C., based on assumed "American" ways of looking at the world, when it is applied to another culture, to people who look at the problem differently based on another set of assumptions and concerns.

What follows is the story of how the Marine Corps—a unique military organization with its own cultural ideals, identity, structure, and ways of conducting business—made sense of a strange "foreign" Washington imperative to conduct culturally effective wars. It is a narrative of the unexpected, sometimes paradoxical solutions that occur when two cultures (even two "American" cultures) collide. And why sometimes "building a well a mile from the village," although inefficient, is actually the most effective way to accomplish the mission.

CULTURE IN CONFLICT

I let my M16A4 do my talking. [We need] more combat training and less touchy-feely crap. A basic cultural brief indicating do's and do not's with basic language sets is about as much time as I am willing to waste.
—*Marine Corps Gunnery Sergeant and Infantry Unit Leader*

Understanding the culture is what makes us money on the ground. It goes without saying that people fear the Marine Corps or worse, hate the Marines, for being in their country. Our ability to have the smallest possible impact on daily life and their culture helps them to open up and trust us. When they trust us the biggest battle of all has been won.
—*Marine Sergeant and Machine Gunner*

Communication and knowledge are key to the success of any campaign. Knowing your surroundings and getting the populace to work with you instead of against you seems to be the most effective way to win the wars of today and tomorrow.
—*Marine Staff Sergeant and Infantry Unit Leader*

The problem is not specifically that we need to understand the cultures of those countries we are occupying. The problem is the Marine Corps' culture. We are not designed as an occupying force and are having trouble adapting to the concept. The easiest solution would be to get the U.S. Army back into the occupying force business so that the Marine Corps can return to its killing/destroying business.
—*Marine Staff Sergeant and Engineering Equipment Mechanic*

Introduction

IN 2005, IN RESPONSE TO THE POLITICAL, CULTURAL, and military challenges facing the United States in Iraq and Afghanistan, General James Mattis[1] established a radical new Marine Corps cultural initiative. The goal was simple: teach Marines to interact successfully with the local population in areas of conflict. The implications, however, were anything but simple: transform an elite Spartan military culture founded on the principles of "locate, close with and destroy the enemy"[2] into a "culturally savvy" Marine Corps. Yet how does one create a Marine as equally capable of sitting down crossed-legged and drinking tea with Iraqi sheikhs as seizing the famed island of Tarawa in a bloody and brutal fight?

This book examines the seven-year trajectory of the Marine Corps' efforts to institute a radical, irregular warfare–focused culture policy into a military organization that views itself as structured and trained to fight big "conventional" wars. It is a tale about the internal cultural wars fought over whether and how the Marine Corps should *"return to its killing/destroying business"* (as the one staff sergeant states above) versus a Corps where *"Knowing your surroundings and getting the populace to work with you instead of against you seems to be the most effective way to win the wars of today and tomorrow."*

In contrast to studies of policy or organizational change, which focus on the policy-*making* organization(s) and/or leadership, this book examines the policy-*receiving* organization and the way that individuals within that organization make sense of the policy and adapt accordingly. The result is

1

FIGURE I.1 Top-down view of policy implementation

a "bottom-up" rather than "top-down" approach to understanding organizational change and policy implementation.[3] Centering my analysis on the culture and identity of the Marine Corps, I examine the internal processes of interpreting, developing, and reworking Department of Defense (DoD) and governmental culture policies to fit within Marine Corps ideals and ways of thinking about and doing business.

The following pages present the challenges and issues of military adaptation and change from the perspectives of Marines from all ranks across the Corps, using their words, their stories, and their experiences: employing an "emic" (internal or native) rather than an "etic" (external or outsider) interpretation.[4] These emic perspectives are based on data from both a six-year longitudinal and qualitative anthropological study of Marine Corps culture training and education programs combined with the quantitative results of an online survey of 2,406 Marines regarding culture and language training and skills in the Corps. Integrating field data from over eighty in-depth interviews, field observations at five Marine Corps bases and nine schools, policy and doctrinal documents, and the statistical results from the survey, I examine how Marines across the Corps (from recruits to general officers, drill instructors to curriculum writers, aviation mechanics to intelligence officers, and those deployed from Afghanistan to Iraq, the Philippines, Japan, Somalia, or Columbia) have sought to incorporate and reconcile two seemingly incompatible identities—fearless warrior and culturally savvy Marine—into training, education, and operations.

This analysis portrays the Corps (and its Marines) as an organization with an independent cultural identity and will—separate and unique from that of the larger Department of Defense, the U.S. government, or, in some cases, even its senior leadership (including the Commandant). I argue that the Corps is not simply a passive recipient of external policy directed from above (Figure I.1); rather, individuals and groups within the Corps interpret, imagine, redefine, and shape policy directives to fit within the norms, processes, structures, and cultural ideals shared within the organization (Figure I.2).

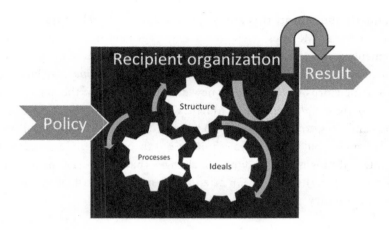

FIGURE I.2 Bottom-up view of policy implementation

As a result, external policies may have entered various Marine Corps organizations in one form, but they "exited" the organization reworked into new forms that fit within the cultural ideals of the Corps. To use a colorful Marine Corps term, the new culture policy became "*Marinized*"—becoming steeped long enough in Marine Corps ways of doing things that it "looks, smells, and tastes Marine."

This process of adaptation and incorporation of the strange "foreign" culture policy into Marine Corps ways of thinking and doing occurred through four unique, though contiguous, methods. As I describe in detail in the chapters that follow, I have termed these four methods: *simplification, translation, processing,* and *reshaping.* Rather than simply rejecting or directly obeying the new external policy directives, Marines and their organizations have used these various methods to adapt and integrate elements of the "foreign" and seemingly incongruous culture policy into Marine Corps ways of acting and seeing the world. The result is a fascinating synthesis of old and new, akin in many ways to the mixing of indigenous and foreign religious elements in syncretic movements.

While this book is based on replicable and comparative scientific field and statistical research methodologies, as Clifford Geertz argues, good ethnographies should provide "thick rich description,"[5] creating a full picture of the world of the subjects in the study by describing intimate complete details of their experiences, rather than painting a thin generalized overall picture. Thus this analysis presents the people in the study using their own words and the world they see through their own eyes.

Equally important, this study takes a culturally relativistic stance, portraying the Marines' viewpoints and experiences without offering judgment, policy, or other evaluations, or any other form of program recommendations. The intention here is *not to evaluate or measure* how well the Marine Corps has adapted to new cultural aspects of the battlespace, but *to analyze and explain why* Marines have interpreted and developed policy and programs as they have. The goal of the study is not to provide an assessment, but to develop a scholarly and theoretical understanding of the internal processes and dynamic changes that occur as a military (or theoretically any) organization attempts to make sense of, adapt to, and implement policy change.

Learning "Marine Speak": Language as a Window into Marine Corps Culture

Perhaps one of the most fascinating aspects of the Marine Corps is its dual identity as an American military organization—with a codified structure and specific governmental and organizational purpose—and as a unique, separate subculture within the United States.[6] Like other subcultures embedded within American society, Marines have a distinct identity, language, history, set of traditions and rituals, and even belief system. However, like all other U.S. government and military organizations, the Marine Corps also has a deliberately created structure and function that must respond to the demands of the secretary of defense and the U.S. government. As a result, the Marine Corps as an institution *deliberately and consciously* reflects upon its cultural ideals and practices—producing mountains of written and electronic materials that promote, advertise, and teach the official Marine Corps culture to civilians and Marines alike.

What is fascinating about this officially created cultural identity—the poster board slogans; the famed quotes of former Marines from commandant to lance corporal; the stark no-nonsense advertising images; the Combat Camera video clips and the eye-catching phrases—however, is that the slogans and images "are not just a bumper sticker," as Major Wagner explained to me in an interview. In fact, many of these classic Marine sayings, and the values and identity that underlie them, have become a cornerstone of Marine Corps folklore, language, and identity, frequently repeated in day-to-day conversations to reinforce and teach basic Marine concepts and beliefs.

As I illustrate in the following chapters, one excellent way to understand and describe Marines is to let them speak in their own language. In fact, Marines have a term for their own unique dialect of English: "Marine speak." These sayings are not simply trite phrases but a window into the Marine Corps way of doing things and its approaches to thinking about and solving problems. For it is through their sayings that Marines teach and model a distinct history and identity that influence how Marines interpret and respond both to external policy and the new aspects of the battlespace.

When I first began my research with the Marine Corps, I viewed these colorful sayings as simply quaint Marine Corps language—propaganda designed to sell a certain image to the public. However, as my research continued, I began to realize how Marine Corps sayings and popular phrases were a cornerstone of Marine self-image: used to teach a moral, to affirm a shared identity, and even to admonish fellow Marines who were not living up to core ideals. In daily speech, Marines would frequently spout out a saying meant to remind their colleagues of a particular lesson or value that needed reinforcing. This cultural norm of expressing emotions and experiences through socially understood sayings is reminiscent of Lila Abu Lughod's[7] analysis of the use of poetry by Bedouin women. Marine sayings, like Bedouin poetry, become a form of culturally understood indirect speech in which the shared stories and meanings behind the sayings or poems communicate concepts or feelings that are unacceptable to discuss in public but completely understood among the speakers.

Due to cultural ideals of toughness, endurance, and invulnerability to pain, Marines do not publicly demonstrate or accept expressions of pain, distress, or frustration. Thus a common response to a fellow Marine who is complaining or expressing difficulty with an issue is not to provide a lecture or (worse) empathize, but to simply end the conversation with a well-understood phrase or saying that illustrates and reinforces the qualities and characteristics that the Marine ought to be demonstrating. For example, a Marine might state, "There's no 'I' in team" (everyone has to work together) to a subordinate who is expressing frustration about another unit that is being uncooperative. No more explanation would be necessary. Rather than directly chide a fellow "Devil Dog" who is waxing too long on a subject, a Marine will respond, "Give it to me Barney style" (give me a simple explanation), or state abruptly, "Where's the 'So What'?" (get to the main point), or "Give me the 80 percent solution" (don't be a perfectionist), emphasizing the need for simplicity and speed in their work. In discussing an unexpected assignment far outside of one's job description—for example, requiring an infantryman to take on the financial management of the unit's project—rather than offering sympathy, a fellow "Leatherneck" will typically respond, "Every Marine a rifleman" (all Marines are capable of the same work; essentially interchangeable) or "semper gumby" (always flexible), and both Marines would then simply shrug and move on to their work.

One of the most interesting words in the entire Marine Corps vocabulary is the verb to "Marinize" something, which means, as Colonel Irons explained in an interview, "to make it look, sound, smell, and taste Marine." This word, above all else, describes the process that I observed of transforming the new policy of creating culturally competent Marines into something that would fit within Marine Corps norms, ideals, and organizational structures. In the following chapters, I describe this transformation over the six years of my fieldwork. Each chapter is centered around a Marine Corps cultural saying that illustrates core cultural values and Marine Corps ways of doing things. Through these sayings, and the Marines' explanations of the way they see the world, the reader can make sense of each of the steps used to "Marinize" culture policy.

To understand and explain the Marine Corps process of "Marinizing" the new external culture policies into existing Marine ideals and ways of seeing the world, I have divided the book into two parts. Part I, "Ideals," describes and explains Marine Corps culture, ideals, and organization from the perspectives of the many Marines whom I interviewed and observed during my six years of fieldwork. Part II, "Realities," then examines the everyday, gritty reality of implementing a culture policy that requires building skills in negotiation, cross-cultural understanding, and even "tea drinking" into this ideal expeditionary Spartan warrior culture.

Setting the stage for this analysis, chapter 1, "When the Boots Hit the Ground," describes the theoretical and methodological basis of the study. Part I, "Ideals," next begins with chapter 2, "Every Marine a Rifleman." Basing the chapter on this oft-repeated Marine Corps saying, I explain the fundamental philosophy in the Corps that Marines are not individuals but part of a Spartan, selfless, flexible, interchangeable team. Chapter 3, "Soldiers of the Sea," describes Marine Corps history, structure, and function, revealing how the expeditionary nature of the Corps and its close maritime ties to the Navy are reflected in the Corps' focus on mission accomplishment and a resulting flexible set of structures. Chapter 4, "Honor, Courage, and Commitment," follows the process of inculcating these values through training at the recruit depots. Finally, "Tip of the Spear," chapter 5, examines how Marines' self-concept as leaders (both personally and as a Corps) is developed through officer training and education, creating and selecting officers (both commissioned and non-commissioned) who exemplify the Corps' values of selflessness, sacrifice, flexibility, decisiveness, and bias for action.

Part II then follows the Marine Corps' efforts to implement culture policy—in theater, in planning, in training and education, and through its manpower system—using four overlapping methods: *simplification, translation, processing, and reshaping.* As I argue in chapter 6, "Building the Plane as We're Flying It," Marines first responded to the need for cultural skills in theater by using the method of *simplification*—viewing the early combat years in Iraq not as something radically different or unusual, but rather as simply "doing what Marines always do." Thus initially Marines rapidly and flexibly adapted to the situation, relying on their "semper gumby," can-do attitude and hiring interpreters and native speakers to "decode" what was unfamiliar. This initial sense of success, however, was soon replaced by frustration as it became apparent that cultural understanding was dramatically more complex, demanding an entirely new way of thinking about the problems in the battlespace. In chapter 7, "The 80 Percent Solution," I illustrate how Marines turned to subject matter experts (SMEs) to analyze the situation and then "*translate*" their knowledge into familiar military ways of analyzing information through graphics, PowerPoint presentations, and computerized simulation programs. The resulting difficulties in speaking across military and academic languages, I argue, illustrated that cross-cultural miscommunication was not simply limited to working with foreign cultures. Chapter 8, "Where's the 'So What'?," discusses how the Marine Corps Center for Advanced Operational Culture Learning (CAOCL) focused on embedding culture and language training into accepted Marine Corps *processes*, turning the unfamiliar, "squishy," multidimensional concept of culture into a strangely restructured, familiar, linear military training format. Finally, in chapter 9, "There's No 'I' in Team," I illustrate how the Corps *reshaped* original government policy mandates to create culture, language, and regional specialists, paradoxically developing, in response, a *nonspecialist* regional, culture, and language familiarization program. This program fit within the existing Marine Corps organizational structures based on a manpower system designed to move "faces" into "spaces" rather than specialists into niches. By "*Marinizing,*" or creating solutions that truly "looked, smelled, and tasted" Marine, the Corps has been able to assign a cultural and regional specialization to every career Marine without either having to change its fundamental ideal of "every Marine a rifleman" or significantly shifting its manpower system.

By shifting research on the military from a "top-down," outside-in approach emphasizing the national strategic level (with its focus on political/ military leadership) to a "bottom-up" inside-out organizational and internal cultural perspective, this study forces scholars to rethink the degree to which national policy makers and military leaders are the only (or even primary) influence on military institutions and their cultures. Using an anthropological perspective, this study depicts militaries as active participants in translating, redefining, and adapting external policies and directives to fit within their cultural and organizational ideals and ways of doing business. The result is a multifaceted analysis that forces readers to reevaluate their assumptions about the U.S. military, the process of policy implementation, and the role of organizational culture and identity in influencing organizational change. Thus ultimately this book is a cultural case study: both of the Marine Corps and of the challenges of its shifting military identity in a new world of irregular warfare.

1   "When the Boots Hit the Ground"

Studying Military Culture from the Ground Up

"One of the hardest things you have to do [as a Marine] is you have to go from being the combatant to the humanitarian worker," began Staff Sergeant Kukela. I am sitting in a stark wooden room on the Marine Corps base at Camp LeJeune interviewing a team of three staff sergeants about their challenges in training their Marines for the cultural aspects of the battalion's upcoming deployment to Iraq. We are seated at a rudimentary table surrounded by four wooden chairs in Battalion 2/6's staff room.

"You really have to switch," he continued. "[For example], I was at an entry control point. We had a vehicle that was on a pullout list, possibly a suicide vehicle. And it switched out of the search area and was driving toward the post [the entry to the base]—the same vehicle as on the pullout list. We ended up shooting—did an EOF [escalation of force] on the vehicle and shot the driver."

But, he explained, as they checked inside the vehicle it became obvious that the driver had no weapons. The staff sergeant leaned forward intently as he described the scene, "[So in less than a second], we went from being that combatant trying to defend yourself to now this guy is just a civilian being stupid. And now we're dealing with a guy who has just been shot."

Staff Sergeant Kukela frowns, "Yeah, so now we've got to save this guy's life. We just shot the guy because he was stupid and all he was being was a dumb driver. Now we have to switch roles and just totally change. Now we've got to save his life and if he had a family with him, we have to deal with his family now. He wasn't a combatant.

"If he was a combatant and you shot him, he's dead. Life goes on. But now you've got to—whoever did that has to deal with the fact that they just shot an innocent person because he was driving dumb. There's a big switch there sometimes . . . the way the fight is going, the way the war is going and what we have to deal with."

To his right, Staff Sergeant Benson leans forward to explain the challenges of contemporary combat, of being a twenty or twenty-two-year-old Marine who is struggling to make sense of a strange new world where people speak incomprehensible languages in a culture that is radically different from theirs. Where simple hand gestures or actions can be mis-interpreted with deadly results. A world where the rules are unclear, and switching from the role of warrior to emergency aid worker can happen in a second. **"Hindsight is twenty-twenty. And anyone doing an investigation, they were not there. They cannot see in the situation that the Marine felt this way, and that is why he did it."**

THIS BOOK IS ABOUT LOOKING AT THE WORLD FROM the Marines' perspective. It is not about whether, at the end of the day, what the Marines did was right or wrong or the best way. For the theoretical question for me, the ethnographer, is not *what* Marines do, but *why.* The goal of this study, as Staff Sergeant Benson so simply states, is to "see in the situation that the Marine felt this way, and that is why he did it."

Most studies of the military, written by those who are sitting on the outside—in academia, government, and policy-making institutions—tend to presume that those service members who live "on the inside" of the military look at and interpret the world the same way as the external civilian researcher. Since the military members come from the same society, the assumption is that while there may be a few occupational or organizational differences, ultimately, civilians and military service members share the same ideals, view and solve problems in a similar manner, and speak and behave in mutually comprehensible ways.

In the case of the U.S. Marine Corps, nothing could be farther from the truth. As I discuss in the methods section of this chapter, I suffered from intense culture shock upon my arrival as the new professor of operational culture at Marine Corps University (MCU) in 2006. Indeed, the experience was all the more exaggerated since I had assumed that I was simply taking another job as a professor at a military-flavored U.S. university. Consequently, while I had mentally prepared myself for the many cross-cultural and linguistic challenges typical of anthropological fieldwork during my previous research in North Africa, in Paris, and among the Navajo,[1] I had absolutely no theoretical or mental preparation for the cultural challenges I would face while working with Marines. This study, in fact, was initially born from my obvious ignorance of Marine Corps culture and my urgent need to understand the way that Marines saw the world in order to communicate and teach effectively in the classroom.

Not only do the majority of studies of the military presume that civilians and military members from the same country share the same culture, but a large body of both political science and sociological research on the military *explains* changes in the military by looking at parallel changes in society and/or the state. In these studies, the dominant paradigm is of a "top-down" "outside-in" model of military change, in which change flows from the state and society to the military (but not vice versa).

However, as I discuss in the following pages, recently a small but growing body of literature, deriving primarily from organizational studies approaches to the military (with a smattering of anthropological research), provides a much more complex picture of militaries. These studies suggest that militaries are not monolithic, warlike mini photocopies of society or the state. Instead, even within the same country, different military services develop their own unique cultures, based on their diverse histories and operational experiences and their differing functions and technology. Furthermore, recent research indicates that military organizations themselves are not homogeneous: differences exist between those service members and units "in garrison" (at home) versus those who are deployed overseas; between officers and enlisted members; and between servicemen and women of different races, gender, or sexual preference. Most intriguing of all are the handful of studies that reveal variations between individual military units within the same service. This research suggests that across the same military service, different sectors may have separate subcultures, due in part to their differing histories and experiences. Several authors point to these findings to argue that military organizations are dynamic "learning" organizations—adapting and changing at the unit level from the "bottom up" as well as from the "top down."[2]

As I discuss in this chapter, these new micro- and organizational level studies of the military form the basis for the methodological and theoretical approaches used in this study.

Military Culture, Organizational Change, and Policy Implementation

Over the past two decades, interest in and research on military culture and change has blossomed across a number of disciplines. Political science and historical perspectives have offered strategic macrolevel studies of national military cultures and the international and political implications of changing state policy and national "ways of war." Sociological and psychological (and occasionally, anthropological) research, on the other hand, has provided a more microlevel approach: focusing on military groups or individuals and the impact of social and policy changes on the sociopsychological and internal dimensions of military life and culture. A third literature has developed as an increasing number of researchers from across these disciplines have incorporated principles from organizational studies in their research. These organizational studies provide several in-depth, midlevel analyses of military organization and structure and the influence of organizational culture on the military's capacity to adapt.

Perspectives from Political Science and History: Strategic Culture, "Ways of War," and the Revolution in Military Affairs

In political science and history, a large and growing body of research has developed over the past twenty years revolving around the sources of military innovation, adaptation, and change.[3] One concept foundational to many of these analyses of military change is the notion that strategic culture—a particular state's a priori approach to thinking about and conducting war and its international relations with other states—is intimately related to its resulting military culture. While various scholars in international relations and political science posit different sources for this strategic culture (ranging from a state's history, its strategic alliances and relations, its experiences in earlier conflicts, or the underlying national culture of the state), few would dispute that states and their leaders influence the ways that wars are fought. The ability of militaries to adapt and change, by corollary, is thus viewed as a product and reflection of national ideals, a state's history (and especially its strategic alliances and military history), and the perceptions and influence of members of a strategic elite.[4] Thus, for example, Adamsky[5] examines how the different strategic cultures of Russia, the United States, and Israel influenced the ability of their respective militaries to adapt to new communication technologies; Cassidy[6] argues similarly for the influence of national history and operational experience on the adaptation of Russia, the United States, and the British to counterinsurgency; and the many articles in Farrell and Terriff's volume *The Sources of Military Change*[7] focus on the influence that national cultural norms, politics, strategy, and new technology have had upon the ability of militaries in different countries to adapt over the past century.

Paralleling this "top-down" approach examining national and strategic influences on military change is the dominant focus of a large portion of policy studies research on the role of government policy makers and external interest groups in influencing change through policy making, policy analysis, and the policy process. A simple examination of popular contemporary introductory texts on public policy[8] reveals a strong emphasis on studying the roles of formal and informal policy actors in creating policy with a much more limited examination of the organizations and groups receiving the policy. Reflecting this perspective, one of the most influential texts on national security and the policy process[9] observes that the military, because of its command and control structure and apolitical position in government, has little to no role in making policy, falling under the authority and jurisdiction of Congress and the U.S. president.

To the credit of the field of policy studies, and influenced in part by research in public administration and management,[10] as well as contributions from sociologists, and, more recently, anthropologists,[11] a secondary literature has evolved focusing on the challenges of policy implementation and the need to understand the perspectives of the recipient organizations and players. Inspired in part by Pressman and Wildavsky's landmark book *Implementation: How Great Expectations in Washington Are Dashed in Oakland*,[12] studies of policy implementation have provided a more "bottom-up" perspective of what actually happens to policies as they reach the organizations and groups that must apply them.

Implementation research has forced scholars to look at the local political dynamics, multiple loci, and organizational networks influencing implementation as well as the challenges of realistically assessing policy outcomes.[13] However, although many of these studies reveal the complexity and dynamics of the various players involved in policy implementation, the specific organizations implementing the policies are often still treated as "black boxes." In contrast, a handful of good and growing microstudies of these organizations (see, for example, Lipsky's *Street Level Bureaucracy*,[14] Yanow's *How Does a Policy Mean?*,[15] and Ingram's[16] research on equal opportunity and the clothing industry in Ireland) suggest that the implementing organizations are neither internally coherent nor necessarily consistent; that the social and economic realities of work in these organizations may undermine the ability of workers to achieve the ideals of the organization; and that these organizations may have their own culturally biased interpretations of the intent of the policy. Such studies bring into question the assumption that implementing organizations are merely inert recipients of policy.

Cultural Variation Within and Between Militaries: Psychological and Sociological Perspectives

In contrast to the strategic culture approach in political science and history, which emphasizes the state and international actors, sociologists and psychologists have tended to focus their research on the internal variations within the military, processes of socialization and internalization of military roles and values, and the effects of military life and combat on individuals and military families and communities.[17] These microlevel approaches provide a view of a military that is far from a homogenous "black box." Studies on diversity within the military illustrate the ways that such demographic and biological factors as gender,[18] race and ethnicity,[19] and sexual identity[20] influence career progression, retention, acceptance within the military community, and even motivations for hazing.

In addition to examining demographic variations in military organizations, sociologists and psychologists have also studied how occupational, organizational, and experiential differences among units can create internal variations within the military. For example, scholars have described the differences between deployed versus garrison cultures in the military (a topic I touch upon in chapter 3).[21] They also note the different branch and occupational cultures that arise in some militaries, such as in the U.S. Army.[22]

On the other end of the spectrum, using an international comparative approach, a number of sociologists and psychologists have also examined cross-cultural differences across military cultures around the world. Reicht[23] compares differing approaches to discipline across militaries from eighteen nations; Labuc[24] compares military training between Soviet, Israeli, and British soldiers; and Soeters et al.[25] compare North Atlantic Treaty Organization (NATO) militaries using the psychological values model of Hofstede. Like the political science approaches, these studies explain difference between military cultures in terms of the differing national character and values of the various countries from which the militaries originate.

Extending this approach, Moskos and his colleagues have published numerous studies examining changes in Western militaries over the past century, relating these changes to shifts in the values of the surrounding culture.[26] In contrast to the state-focused studies of political science, this research examines the internal cultural and organizational shifts within militaries as the result of parallel shifts in society: positing a postmodern shift in Western militaries from a more institutional to a more professional occupational identity and organization. However, both disciplinary approaches agree that the cause of these changes is due to external cultural shifts—whether "top-down" shifts in the nation-state and strategic policy or "outside-in" changes in the values, ideals, and orientations of the general public and society from which military members originate. As Moskos states, "The military can be understood as a social organization which maintains levels of autonomy while refracting broader societal trends."[27]

In general, both political science and sociological/psychological approaches focus on the ways that military organizations are a reflection of or natural extension of the state and society. In such models, militaries are not viewed as separate or distinct from the state or society. Since such models presume that the state and society dictate military culture, changes in military culture, accordingly, are explained as the result of larger changes in the state or society. However, a small but growing number of organizational and "bottom-up" studies of militaries suggest that, in fact, military services have their own distinct histories, identities, and cultures that also may significantly influence the ability of a specific service or branch to adapt and change independent of the surrounding society or state.

Disappointingly, the discipline that could contribute the most to our understanding of military change at the microlevel has, for the most part, completely avoided research on militaries as cultures.

Anthropology and the Military

Unfortunately, anthropological approaches, which focus on a native, or "emic," perspective and emphasize the internal dynamics and processes within cultures, organizations, and groups, have largely been absent from studies of the military. This gap is due in part to ethical concerns and in part to a historically fraught relationship with the military.[28] However, it also reflects a bias in anthropology toward studying disempowered minorities and exotic cultures. The result is that American anthropologists have tended to avoid research on the military directly, focusing most of their research *around and about* the military, rather than conducting studies *of* the military.

Despite Laura Nader's influential article (written in 1972) on the importance of "studying up" to powerful institutions,[29] in the field of anthropology the preference still remains to "study down" and focus on disempowered minorities and foreign cultures.[30] This bias toward studying the disenfranchised rather than the powerful institutions and groups that can influence the lives of those "at the bottom," however, means that in-depth, ethnographic research on such nationally and internationally influential institutions as the military (and larger security sector), governmental organizations, and international corporations is patently lacking.

Indeed, the absence of anthropological research on the world of powerful decision makers has meant that, increasingly, the discipline has become marginalized and irrelevant in governmental and policy-making spheres. Reflecting his concern with the lack of anthropological involvement in national policy decision making processes, in his book *Peacekeeping under Fire*, Rubinstein defends his unusual choice (for an anthropologist) to study peacekeepers by stating, "I was deeply concerned that anthropological knowledge was not being used in a meaningful way by policymakers. . . . I was especially concerned about the lack of anthropological perspectives in U.S. national security policy."[31]

Further complicating the issue, historical and contemporary political hostility toward the military on the part of many American anthropologists[32] has made it professionally awkward, if not career suicide, to study or work with actors in what Albro et al. term the "securityscape."[33] Consequently, in American anthropology, the study of militaries as unique cultures and organizations has been generally neglected, with a few notable exceptions. Considering that anthropology is a discipline that prides itself on participant observation and fieldwork—living with and sharing the experiences and the lives of the people in the study—the resulting studies of war, the military, and security organizations are a rather strangely distant and almost disembodied set of writings *around* (but not touching) the military and the associated security apparatus.[34] Thus anthropologists have produced an extensive literature on the anthropology of war and conflict,[35] a number of studies that examine the interface of militaries with civilian populations (ranging from studies of the impact of military bases on host communities to the experiences of military wives),[36] and even several studies that examine the military and its associated security apparatus from the *outside*—using military publications, doctrine, correspondence, and advertisements as their sources of ethnographic data rather than actually interviewing or observing military servicemen and women and their activities directly.[37]

Interestingly, in contrast to American anthropologists, European, Latin American, and Middle Eastern anthropologists have been actively engaged for some time in direct fieldwork among militaries in their countries, suggesting that the notion of the military as a "ritually polluting"[38] subject of study may be a uniquely American concern. Thus several Latin American anthropologists have written ethnographic studies on Brazilian and Argentinean military organization and hierarchy.[39] Kirke, a British anthropologist, has provided research on the social structures within the British army.[40] And the Israeli anthropologist Ayal Ben Ari has devoted his career to various studies of life in Israeli and Japanese military units, examining the conflicting ways that military symbols, language, and ideals are understood, interpreted, and enacted by the men and women who work and fight for the Israeli Defense Forces (IDF) and the Japanese army.[41]

Despite the biases within the American anthropological community, in recent years a handful of American ethnographies of military and security organizations have begun to add to our understanding of the internal dynamics of military units and groups. In addition to Rubinstein's studies of peacekeeping organizations,[42] Hawkins examines the internal and external contradictions of an Army community deployed as a NATO force in Germany, and Simons explores the internal dynamics of an Army Special Forces unit. Each of these studies illustrates the complex interplay between military units and external (political and social) forces as well as the internally complex relationships within military units.

Several other ethnographies have focused on the symbolic and linguistic dimensions of security institutions and military life. Tortorello[43] explores how the Marine Corps value of courage becomes symbolically embodied through the process of martial arts training. Burke[44] describes the way that folklore, traditions, and language reflect and remake a culture of masculinity in the Navy. And Gusterson[45] analyzes how rituals and symbols are used to construct certain culturally accepted "truths" about nuclear weapons among nuclear scientists in the Lawrence Livermore International Laboratory. These studies emphasize the important roles that symbols, rituals, traditions, and language play in reinforcing and re-creating military identities and values.

Finally, research by Fosher and Fujimura expands our understanding of militaries and security institutions to examine how these cultures interpret and respond to change. Fujimura's research[46] on midshipmen and curricula at the U.S. Naval Academy explores the conflicting and changing views of "soft cultural studies" at the academy in response to new political imperatives during the past decades. In contrast to Fujimura's more historical perspective, Fosher's study focuses upon the dynamic and changing homeland security community centered around the Boston area.[47] Her research emphasizes the way that shifting, vague, and contradictory government policies were interpreted and "constructed" among various members of this community, emphasizing the process rather than the outcome of policy interpretation and implementation.

Each of these anthropological studies portrays the military not as a monolithic institution, but as a set of dynamic groups or communities that actively negotiate and interact with each other, seeking to interpret and adapt to often conflicting external and internal imperatives and situations. In this approach, individuals and groups within the military are seen as active agents, rather than passive participants in the process of change and adaptation.

Organizational Culture and Military Change

While anthropologists have provided a handful of intriguing views of the military from the inside, much of our understanding about the internal dynamics and nature of individual militaries comes from the use of the cross-disciplinary approaches of organizational studies. A number of scholars from across the fields of political science, sociology, and anthropology have drawn upon organizational studies theories to refine their analyses of military culture and change, focusing on militaries as organizations. Two dominant approaches to the study of organizational culture have been particularly influential in studies of military culture: the structural approach deriving largely from sociological studies, and the symbolic approach influenced strongly by anthropological methods.[48]

Structural studies of military culture and change tend to focus on military organizational hierarchies, manpower, roles, and branches. Not unexpectedly, a common focus in such studies is upon the role of military leadership in influencing change within the organization.[49] Perhaps the most well known study using this approach for Marines is General Victor Krulak's *First to Fight*,[50] which delineates five different kinds of leaders who have influenced the Corps throughout its history. Terriff[51] also focuses on the role of leadership in effecting (or failing to implement) lasting change in the Marine Corps by comparing the efforts of two significantly different commandants of the Corps: General Krulak and General Gray. In contrast to these "top-down" examinations of military change, Soeters[52] and Pierce[53] examine the structure of the military organization itself, arguing that historically rigid military organizational structures tend to inhibit rapid change.

Symbolic approaches, in contrast, focus on organizational identity, beliefs, and ideals and their expression through symbols, rituals, ceremonies, and (historical) narratives. In four different studies by Terriff,[54] Hoffman,[55] Connable,[56] and Smith,[57] the authors illustrate how the various U.S. military services' history, identity, and narratives have influenced the ability of these services to adapt to contemporary counterinsurgency and cultural challenges. As these studies reveal, by focusing on specific military organizations rather than on a generic "national" military culture, the cultural and organizational differences between military services and their branches become apparent.

In several elegant studies that seek to integrate different theoretical models of organizations, Martin[58] delineates three distinct models of organizational adaptation and change: the integration perspective, the differentiation perspective, and the fragmentation perspective. The integration perspective portrays culture as relatively homogenous within an organization and emphasizes the way that shared values and ideals unite organizations and differentiate them from other organizations. Using this perspective, Labuc,[59] Soeters et al.,[60] and English[61] each argue that the national cultures of the different militaries in their studies influenced their internal structure, values, and modes of operations. These differences in turn, they argue, affected the ability of the various militaries to adapt to contemporary crises.

The differentiation perspective, on the other hand, portrays organizations as composed of varied and competing groups that can be considered more of an actively debating "political community"—a portrayal that fits the U.S. military with its interservice rivalry and strong service cultures quite well. Mahnken,[62] for example, compares the ability of different U.S. services to adapt to new technology over the past fifty years, relating their different responses to the unique culture of each service.

Both the integration and differentiation perspectives do assume some coherent shared cultural identity and norms—the issue being the level of coherence (across the organization or within subsections of the organization). Fragmentation perspectives, on the other hand, emphasize ambiguity and lack of consensus within the organization—suggesting that there may be multiple and contradictory interpretations of what the organization is doing and why. Reflecting this perspective, the collection of essays in Rosen,[63] for example, examines the internal dialogues and debates within specific United States and British services during periods of peacetime and the influence that such dissension had upon the abilities of the specific services to change. Some exciting variations on the fragmentation approach are the studies by Mahnken, Nagl, and Russell,[64] which view militaries as internally organic, adaptive, and learning organizations. These researchers provide strong evidence that military adaptation and change can and does occur unequally across organizations. Russell's research in particular provides compelling evidence that during the Iraq war, differing Army and Marine units adapted to new realities on the ground, making tactical and operational changes far in advance of guidance from leadership at the top levels.

In this study, I employ each of these organizational perspectives at different points in the analysis in order to explain the ways that Marines in the Corps adapted to the new cultural realities of the battlespace. Part I offers an idealized symbolic and integration perspective: describing an ideal coherent Marine Corps identity, based on a shared history and identity, an integrated training program focusing on values, and the selection of officers with ideal Marine qualities. Part II examines military change through the lenses of the differentiation and fragmentation perspectives as Marines in my study explain the personal and organizational challenges of responding to the new cultural realities of irregular warfare in Iraq and then Afghanistan.

Like the Marines' early ad hoc, creative, and improvised solutions to incorporate cultural understanding into military operations, my own trajectory in designing and conducting a study of the Marine Corps followed a similarly unusual and unexpected path.

Learning by Doing: Researching Culture by Teaching Marines

In January 2006, three years after the initial invasion of Iraq, the Marine Corps Center for Advanced Operational Culture Learning (CAOCL) received its official charter from General James Mattis (at the time, Lieutenant General, commanding Marine Corps Combat Development Command). The center's initial purpose was to provide culture and language training for Marines deploying to Iraq. However, fairly quickly the center's leadership realized that teaching a few phrases in Arabic along with some basic courtesies (or "do's and don'ts" as the Marines called it) was not sufficient to provide the in-depth cultural understanding required to solve the problems the United States was facing in the new irregular warfare environment. So in collaboration with MCU, the Corps' nationally accredited graduate university for its senior military officers, CAOCL funded a new and unique position: a "professor of operational culture" who would teach graduate-level courses on anthropology, Islam, and the Middle East.

On September 11, 2006, in the midst of some serious virtual scholarly tomato throwing at any anthropologist who would have the nerve to actually work with the military,[65] I began my first day as the professor of operational culture at MCU. From the beginning, I faced two serious problems in this new position (neither of which had to do with the debates in anthropology about the ethics of working with the military).[66] First, I had no idea what *operational* culture was—and nor for that matter did the Marine Corps. In fact, one of my first tasks in my job was to define the concept, an interesting initiation into the convoluted pathways that cultural training was already taking in the Marine Corps. Second, as a blonde, female anthropologist standing 5'4½" tall, with no prior experience with the military, I was expected to teach operational culture principles to classrooms of "salty" (seasoned) thirty-to-forty-year-old Marine majors and lieutenant colonels who had already deployed numerous times to Iraq, Afghanistan, Bosnia, the Horn of Africa, Nicaragua, and more. To say that the students and I both faced serious culture shock as we struggled to translate our worlds to each other would be an understatement.[67]

I quickly realized that I needed to design and teach a culture curriculum that would fit the unique needs of my students who saw culture more as a verb (something one *does* to accomplish the mission) rather than a noun (something one observes and studies). And to do that, I would first have to understand the culture of the Marines in my classes. Thus this research initially began as a curriculum development study to understand what Marines did, how they viewed the world, and how that influenced their learning styles (a classic educational ethnography) in order to create classes that were relevant, interesting, and culturally appropriate for my audience. The leadership at Marine Corps Training and Education Command (TECOM), MCU, and CAOCL saw this research as a valuable part of my work and professional development, generously sponsoring the project over its six-year span.[68] There was also some not inconsiderable curiosity on the part of my Marine Corps counterparts regarding what it was an anthropologist actually did—and how she did it. So in many ways the project was born as a mutual opportunity for each of us to learn about the other's methods and ways of interpreting the world: in essence, practicing and developing the cross-cultural understanding that I was teaching in the classroom.

Initially my research focused on the resident Marine Corps officer and enlisted education programs. I not only taught my own classes at the four officer and one staff non-commissioned officer (NCO) schools at the university, but I also devoted significant time to observing classes and exercises taught by other faculty, both civilian and military: a classic participant-observation approach. Furthermore, I observed and interviewed instructors and their classes at the Officer Candidates School (OCS), The Basic School (TBS), and the infantry officer's course (IOC).

Over time, however, the project expanded[69] to focus not only on the ways Marines learned in the classroom but also on the ways that the new culture and language requirements were being interpreted and implemented into Marine Corps organizations and culture—from the recruit depots to the professional military education (PME) classes; from short-term training prior to deployments to programs focused on long-term cultural skill development such as the foreign area officers (FAOs) and the training advisory groups.[70] Thus, in addition to conducting ethnographic observations and interviews at seven of the resident Marine Corps PME schools, I also traveled to the different MEFs (Marine expeditionary forces—see Appendix) on the East and West Coasts. With the support of CAOCL, I observed Marine Corps cultural training programs and predeployment exercises on four different Marine Corps bases around the country: at Camp LeJeune, North Carolina; Marine Corps Air Ground Combat Center (MCAGCC) in 29 Palms, California; Camp Upshur on Marine Corps Base, Quantico, Virginia; and Marine barracks at 8th and I in Washington, D.C. Observations ranged from observing culture and language classes and briefs for battalions or other units preparing for deployment, to watching simulated cultural exercises such the Mojave Viper combined arms exercises (CAXs) in which Marines interacted with live Iraqi (and later Afghan) role players as they moved through a mock Middle Eastern village, to observing Marines conducting rifle training and field maneuvers, to a visit on board the Navy/Marine ships USS *Bataan* and USS *Gunston Hall* and a one-week deployment on the Navy aircraft carrier USS *Nimitz* to observe life at sea. Finally, I spent an intense week of observation and interviews with the leadership, instructors, and recruits at Parris Island Recruit Depot in South Carolina.

To make sense of what I was observing, I also conducted in-depth interviews with over eighty enlisted Marines and officers (both active duty and retired). The interviews spanned all ranks, from second lieutenant to general officer, as well as from recruits at the depots (boot camp) to sergeants major. While most of my interviews centered on the challenges of implementing cultural programs in the United States, a number of Marines who had returned from recent deployments to Iraq and Afghanistan also kindly volunteered to discuss their cultural challenges while operating in the field.[71] Thus the study not only includes data on programmatic challenges and processes but also individual Marines' on-the-ground views of their need for and use of culture "in theater."

The Online CAOCL Culture and Language Survey

Last but not least, I led the design, implementation, and analysis of an online survey by CAOCL focusing on attitudes toward culture and language learning.[72] Launched in February 2010 with the assistance of the Marine Corps Center for Lessons Learned,[73] CAOCL sent out an anonymous survey to 15 percent of all Marines (except general officers) with addresses on the Global Address List (GAL). The survey sample consisted of predominantly career Marines (with ranks of lieutenant or corporal and above): in other words, those Marines who had been in the Corps long enough to have been deployed and to have received culture and language training in the past four years.[74]

We received 2,406 valid responses from active-duty Marines from every rank (except general officer), and every military occupational specialty (MOS). Eighty-three percent of the respondents had deployed at least once during their careers, and 20 percent were currently deployed at the time they filled out the survey. Table 1.1 describes the characteristics of the survey sample. These characteristics are congruent with manpower data for career Marines[75] published by the Marine Corps[76] and indicate that the sample is typical of career Marines in the Corps.

The survey asked Marines basic demographic questions regarding their rank, military occupational specialty, age, gender, education, language and cultural background, and deployment experience. Marines were also requested to answer a series of questions about their use of culture and language skills while deployed, their predeployment culture and language training, the usefulness and value of culture and language training for preparing them for the mission, and how important they thought culture and language skills were, in general, for mission effectiveness. In addition to a set of standardized questions, Marines had several opportunities to provide open-ended answers and suggestions.

Results were analyzed using the statistical package for the social sciences (SPSS) and included both statistical analyses of the quantitative data and qualitative analyses of the open-ended answers. Where appropriate, and with deep thanks to CAOCL, I have also included both the statistical results and qualitative answers from the survey in this book. A complete description of the survey questionnaire, sampling frame, and methodology are available from CAOCL upon request.[77]

The Longitudinal Data

Although the research had not initially been conceptualized as a longitudinal study, due to the many other immediate demands of my job and the constantly evolving issues surrounding culture policy for the Marine Corps, ultimately the research spanned more than six years. Thus I had the unusual opportunity of watching and participating in the Marine Corps' day-to-day challenges of implementing its culture policy requirements through virtually all of its twists and turns. Since an important part of my job was supporting the work of CAOCL, I participated in many of their discussions regarding incorporating cultural and language skills into doctrine and training manuals, as well as the program's efforts to work with other Marine Corps

TABLE 1.1 CAOCL survey: demographic characteristics

Rank N=2,406	Education N=2,406	Ever deployed? N=2,406	Culture training N=1,659*	Language training N=1,642*
76% enlisted	64% High school grad	83% Yes	84% Yes	55% Yes
24% officer	36% College grad	17% No	16% No	45% No

SOURCE: CAOCL survey, 2010
*Percentage is based on those Marines who reported they had deployed anywhere in the world in the past four years.

organizations such as the Foreign Area Officer (FAO) program, the Marine Corps Intelligence Activity (MCIA), the Marine Corps Information Operations Center (MCIOC), the Marine Corps Training and Advisory Group (MCTAG), the Civil Affairs program, and the Manpower and Reserve Affairs (M&RA) department.

Although it was not my primary role, I also was required occasionally to participate in higher level debates and discussions regarding the development of culture policy at the Defense Language Office, which was given the leadership for overseeing and implementing culture policy for all the services. Finally, as each military service struggled with its own efforts to develop its own specific culture and language programs, we formed a small but collegial cross-service group where we shared challenges and successes and learned from each other.

This book, then, is based on a broad, but also deep, six-year longitudinal study of the fascinating, convoluted, sometimes frustrating, and always exciting and unexpected process of incorporating (and reshaping) cultural policy across the Marine Corps: from predeployment training to graduate-level education programs; from recruits in boot camp to battalion commanders returning from Afghanistan and Iraq; from personal interviews and observations to a large-scale statistical study; and from the trailers of CAOCL to the hallowed halls of the Department of Defense.

Ethical Considerations

Ethnographic research, regardless of context, is always fraught with ethical issues. Because the purpose of participant-observation is to fit into the social environment so that actors "behave naturally,"[78] the lines between one's official role in the situation (whether as a professor, researcher, student, employee, or simply weird tourist) and one's personal relationships built in the field (friend, coworker, adopted family member, or even wealthy foreigner with access to unique goods and privileges) are frequently difficult to navigate. For example, as I learned during my earlier fieldwork in Tunisia,[79] while I may have assumed that my invitation to, and attendance at, a wedding was for scientific and scholarly purposes, the bride, groom, and their families and friends might see me as a dear friend and guest, privileged to participate in activities and conversations that could cause them intense pain and embarrassment if published in a book that others could read about them. Sadly, historically, anthropologists were able to "avoid" many of the ethical repercussions of the intimate personal details they published about individuals from cultures far away, since it was assumed that the "subjects" of the study could not read English and would never see what was written about them.[80]

Fortunately, as the world becomes smaller and people even in the most remote locations become literate and gain access to the Internet and other media, most anthropological research today (and especially research that "studies up" in the United States to institutions of power and authority) must weigh carefully the ethical implications and responsibilities resulting from the publication of the anthropologists' work. As Gusterson states, "In the past anthropologists wrote about people who lived far away for audiences that had to take our interpretations largely at face value. The anthropologist who studies up in the United States will have colleagues who may already have strong opinions about his or her subjects, who may even have met them, and subjects who will read and argue with what is written about them."[81]

Contemporary anthropologists, whether working in the Amazon jungle or among English-speaking Marines (who—hopefully—will be reading and debating the results of this study after it is published), must carefully weigh whether comments and actions observed in the field are to be included as publishable data or viewed as a supporting backstory that helps inform the researcher's thinking and understanding. In my case, distinguishing between my work as a professor at MCU and CAOCL and my research for this book required much careful ethical consideration. While the practice of work, as others have noted,[82] allows researchers to participate in the "embodiment" of cultural activities, it also requires a much more thorough examination of the resulting data and its ethical use for publication.

In order to protect participants in the study and to respect their understanding of my different roles in the workplace, I have made a careful distinction between data obtained during official field observations, public events (such as conferences and speeches), and interviews versus those observations resulting from my personal professional internal engagement with those involved with the development of culture policy and curriculum for the Marine Corps. Since most meetings and policy conversations were part of my regular day job and were not included under my research protocol, to protect the actors in these events, the specific internal details of discussions at meetings on culture policy or interagency coordination are omitted from the data presented here. However, my official interviews with the actors are included, as are the publicly released policy documents, doctrine, manuals, and other publications that resulted from these internal meetings. Furthermore, since the Marine Corps openly publishes and writes about itself on a regular basis, I have included formal Marine Corps publications, advertisements, and videos as well as publicly released after action reports (AARs) from Marine Corps Center for Lessons Learned and other sources to support my analyses.

Interestingly, highlighting the Marine Corps values of responsibility and integrity (see chapter 5), my use of pseudonyms, along with changing rank, gender, billet, MOS, or other identifying demographic data (when necessary to protect the identity of speakers), brought some unexpectedly negative reactions from a number of the Marines I interviewed. Several of them believed that masking their identity was "cowardly" and that a Marine should stand by what he or she said. Indeed, at one point I found myself in the bizarre position of having to explain why it was more "honest" and ethical to change the facts regarding one colonel's deployment to Iraq to a different billet in Afghanistan in order to mask his identity. His arguments were compelling, and I ended up changing his history back to the original facts, although it was clear to me that many in the Marine Corps (which is a very small world) could probably determine who the speaker was.

While protecting the identity and respecting the privacy of the participants in this study was a major ethical concern, an equally important concern for readers is probably the intended use and application of my research by the Marine Corps. Curiously, as with my unexpected discovery that, for Marines, masking one's identity was not considered the best ethical treatment of the data, paradoxically, this is probably the first study in my entire academic career that has not been subject to external reporting requirements by the funding agency. I have never been required to provide findings, after action reports, formal or informal policy assessments, or any other official record of my ethnographic research at any point during the six-year research project. (Due to professional courtesy, occasionally I have shared generic verbal and informal observations with the various individuals and organizations that have participated in the study.) However, for the most part, this research has been considered part of my personal "professional" development: the only expected benefit for the Corps was that, with increased understanding of Marines, I would be able to develop culture curriculum better adapted to Marine Corps ways of learning. Likewise, reflecting the university's policy of academic freedom for its faculty, this book has not been subject to official review by any Marine Corps organization and represents my own interpretation and analysis, not the official findings or opinions of the U.S. government or the Marine Corps.[83]

The analysis that follows, then, is my own—not that of the Marine Corps or any other governmental institution. The words and stories that fill the next chapters, however, belong to the Marines who generously offered their time, opened up their schools and classrooms, and shared their personal stories as a legacy and lesson for those who will follow them. This is a study of military change and adaptation, but not from the traditional perspective of those of us on the outside looking in. Rather, it is the story of those on the inside of the Corps who are looking out and thinking, as Staff Sergeant Benson states clearly at the beginning of this chapter, "Hindsight is twenty-twenty. And anyone doing an investigation, they were not there."

1 IDEALS: CORPS CULTURE

STUDYING CULTURAL IDEALS PRESENTS A PARADOX. On the one hand, if we examine any specific population, not all people's behaviors will be congruent with the ideals accepted by the group. Indeed, often only a small percentage of that population will succeed in living up to these ideals. Take, for example, the American ideals of beauty espoused and promoted throughout American media: a slender body, athletic physique (particularly for males), and smooth, unblemished skin, with glossy, thick hair. However, a quick stroll through any shopping mall in the United States will reveal that only a minority of the people shopping there even approximate this ideal—in part simply because of the demographic fact that the majority of Americans are under sixteen and over thirty-five (the age range where such beauty ideals are most easily achieved).

On the other hand, the failure of individuals to achieve a cultural ideal in no way indicates that the ideal is irrelevant, or that the ideal does not permeate and influence every aspect of people's lives. Thus while only a minority of Americans may actually achieve the cultural ideals of beauty, if we examine the U.S. economy and the daily activities of Americans, it is quite clear that an enormous percentage of the population is willing to go to any expense and effort, undergoing even dangerous and painful activities, in order to achieve this ideal of beauty. The sale of cosmetics and beauty products is a multimillion dollar industry. So is the rapidly growing diet and exercise industry, as evidenced by the increasing numbers of gyms and fitness centers, online and personalized diet programs, and even personal trainers and coaches hired to assist Americans in their quest for thinness and fitness. Indeed, Americans take the ideals of beauty and thinness so seriously that they will undergo painful plastic surgery, the insertion of gastric bands, and other bariatric procedures, and even die—starving themselves through anorexia nervosa—to look like the models in the magazines and actors on TV.

Cultural ideals are, in fact, some of the most immensely powerful influences on human behavior, regardless of whether individuals are consciously aware of them or actually succeed in achieving these ideals in any statistically reliable pattern. As the famed French anthropologist Claude Levi-Strauss[1] has argued, cultural ideals are woven into the very fabric of society, *structuring* the way that people think and speak about their lives and experiences. Expanding this concept, Bourdieu[2] speaks of the "habitus," or a set of underlying cultural orientations, that, as Hanks explains, influence one's daily "habits, dispositions to act in certain ways, and schemes of perception that order individual perspectives along socially defined lines."[3]

Part I of this book focuses on these underlying structures of meaning, the ideals and "dispositions to act" that form the Marine Corps "habitus." To understand the Marines' unique responses to external pressures to adapt to the irregular warfare environment in Iraq and Afghanistan, it is first necessary to explain the ideals that influenced how Marines made sense of these new strange external pressures. Thus my goal in the following chapters is not to provide statistical factual data, for example, on how many Marines are actually excellent sharpshooters. Rather, the intent is to explain *why* Marines believe that "every Marine is a rifleman" (chapter 2) regardless of the practical reality of marksmanship in the Corps.

Since language is one of the fundamental structures that reflects, influences, and reproduces cultural ideals, throughout my analysis I have intentionally described the Corps using the words and experiences of Marines.[4] At points where I have included detailed field observations (in particular, at boot camp and the officer training schools), I have taken care to include Marines' narratives about what I was observing. And each chapter is intentionally centered around core Marine sayings, "finding culture in talk" as Naomi Quinn so eloquently describes it.[5]

Part I of this book, then, weaves a story of how Marines see themselves and the Corps. Two of the chapters focus on the creation of Marines at boot camp and the initial officers' schools of OCS and TBS, since these are the locations where Marine Corps culture and ideals are first taught. A third chapter, "Soldiers of the Sea," examines the fascinating relationship between the Marine Corps' organizational structure and Corps ideals of flexibility, illustrating how, in the Marine Corps case, military structure and behavior reflect and adapt to ideals, rather than vice versa. The opening chapter also examines one of the unique and yet most powerful aspects of Marine Corps identity—its egalitarian ethos of an interchangeable body of Marines—an ethos that ultimately has shaped the Corps' responses to external pressures for change in unexpected ways.

This theoretical and methodological approach creates a highly idealized picture of the Corps. And I imagine that at various points a number of the Marine readers of this book will argue that such ideals do not always match daily realities and practice. In a few cases, where a clear tension has been growing—for example, between the Marine's identity as a knuckle-dragging infantryman versus the increasingly technologically sophisticated specialties of many Marines—I provide the counterdialogues. However, for the most part, my purpose in Part I is not to describe the realities of the Corps (the task of the second part of this book) or the many differentiated variations and subgroups of the Corps (which in ideal Corps narrative do not exist) but to explain this ideal narrative. Thus Part I provides what Martin[6] refers to as the "integration perspective" in organizational culture studies (see chapter 1): a view of the Corps as a coherent, unified, and integrated culture with a set of foundational core values that all Marines share.

2 "Every Marine a Rifleman"

The Egalitarian Military Service

ON THE EARLY MORNING OF SUNDAY, MARCH 23, 2003, a hapless U.S. Army convoy rolled along Route 7 headed mistakenly through the eastern suburbs of Nasiriyah on its way to Baghdad, Iraq. The Army's 507th Maintenance Ordnance Company was composed primarily of combat support troops whose function was to provide logistical support to the infantry and other ground troops ahead of it. The convoy included a computer technician, supply clerks, cooks, diesel mechanics, heavy equipment mechanics, and supply sergeants.[1] Included among the soldiers was Private First Class (PFC) Jessica Lynch, a supply clerk.

The convoy passed a group of Marines from Task Force Tarawa. The Marines were in the midst of securing the city, which was still unstable four days after the invasion of Iraq on March 19. Several of the soldiers on the convoy were surprised to be passing combat forces, since their route was planned to bypass the city. In fact, unbeknownst to them, they had lost their way and were headed directly into an ambush.

According to an ABC News report, one of the soldiers became concerned about their route because the convoy was not protected by infantry forces. "We are supposed to enter a town after it has been secured by combat forces," he told the reporter. "Even when an area is completely secure, the maintenance team is still supposed to be protected. They never go anywhere alone."[2]

The exact details of what followed next are still debated to this day. What is known is that the convoy was ambushed. Eleven soldiers ended up dead.

And Jessica Lynch wound up missing, to be recovered in a dramatically broadcast rescue a week later on April 1, 2003. While the details of the ambush and the rescue make a fascinating story for another book, what was important to me as an anthropologist and ethnographer was the meaning that the story held for Marines. Indeed, over the years I have worked for the Marine Corps, the Jessica Lynch ambush has been mentioned numerous times as a primary example of why "every Marine is a rifleman."

"Every Marine a rifleman. It's essential," emphasized Captain Hutchinson, who had been part of Task Force Tarawa in Nasiriya when the Army's 507th Maintenance Ordnance Company had rolled on by. "It's what distinguishes us from everyone else. We picked up the remnants of Jessica Lynch's units. And those soldiers [were completely unprepared]. They didn't think they would have anything to do with fighting."

In the late 1980s, General Alfred Gray, the 29th Commandant of the Marine Corps, spearheaded a campaign to ensure that all Marines had basic combat skills. In contrast to the Army, Navy, and Air Force, which have units and military occupational specialties (MOSs) that are not required to be proficient in combat skills, General Gray wanted a Marine Corps in which every Marine was capable of fighting. The famed quote (popularized by General Gray)[3]—"every Marine is a rifleman"—has become a foundational principle in Marine Corps training and ideology. In practical and training terms, *every* Marine—regardless of whether he or she is a scout sniper, an aviation mechanic, or a supply clerk like the Army's PFC Jessica Lynch—is not only capable of knowing what a rifle looks like, but is trained and requalified (every year) to shoot it effectively in combat. Indeed, so important are marksmanship skills that, according to an exhibit at the National Museum of the Marine Corps in Quantico, Virginia, "If a recruit can't shoot, he can't be a Marine" (attributed to Chief Warrant Officer Anthony Carbonari).

As far as the Marine Corps is concerned, there is no such thing as a "safe" noncombat job in the military. When Marines enlist or take a commission, they (and their parents) are not wooed to the Corps by promises of a solid paying job in protected positions far away from combat. The Jessica Lynch convoy ambush epitomizes to Marines why they must all be riflemen—in combat no one can ever assume that an infantry battalion or a team of scout snipers is hanging around waiting to rescue them in a crisis. As Staff Sergeant Hanson, a drill instructor, noted during an interview, "Even if we have an Admin [administrative support] guy headed to Iraq and even if he is going to be headed to the FOB [forward operating base], he's going to have to take transport from Kuwait to the FOB. The insurgents don't say, 'Well they're an Admin guy, we can't IED them' [blow them up with a homemade bomb]." Thus all Marines are trained and prepared to fight and take care of themselves in combat, even if their assigned job is simply to file papers or maintain airplanes.

This ethos, however, extends far beyond the notion that all Marines should be able to carry and fire a weapon effectively. As I discuss in this chapter, "every Marine is a rifleman" has become the foundation for a culture that emphasizes unity and teamwork and is people—rather than technology—focused. This ideology underlies the Marine Corps' reluctance to develop specialization among its members. In fact, a core value is that no Marine is unique or different or "more special" than any other. Individuality is frowned upon, and all Marines are seen as interchangeable members focused on helping "the guys on the ground" accomplish the mission.

And that quintessential "guy on the ground" is Lance Corporal Binotz.

Lance Corporal Binotz

I do not know whether a real Lance Corporal Binotz ever existed. But in Marine Corps mythology and parlance he is as real, if not more real, than any living or dead Marine. For Lance Corporal Binotz represents the quintessential Marine. He is the young enlisted Marine infantryman, "the guy on the ground"—out there, fighting the fight, to achieve the mission. Lance Corporal Binotz is everyman, or rather "everymarine." He is faceless, not an individual, not someone with a personal history or identity. Interestingly, and a reflection of Marine Corps ideals regarding the relationship between enlisted Marines and officers (see chapter 5), he is not an officer, but a junior enlisted Marine. And he is the sole focus of everyone in the Marine Corps.

This emphasis on a mythical, young, generic lance corporal reflects the actual demographics of the Marine Corps. According to a 2012 publication,[4] "The Marine Corps is the youngest, most junior, and least married of the four [U.S.] military services." In fact, in 2012, 61 percent of the Marines were twenty-five years old or younger.[5] Not only does the Marines Corps have a significantly younger population than other U.S. military services, but in comparison to the other services, the Corps is the least "top heavy" (that is, with a large proportion of senior officers), composed predominantly of junior enlisted Marines. For every Marine officer, there are 7.8 enlisted Marines. In comparison, for every officer in the partner services, there are 4.9 enlisted sailors, 4.5 enlisted soldiers, and 4.1 enlisted airmen.[6] Lance Corporal Binotz thus represents a Marine Corps that values and focuses on its young junior enlisted members, not its senior officers.

"Lance Corporal Binotz—he is the reason we are here," stated Colonel Thompson, who had recently left command of a Joint Task Force to return to Marine Corps Base Quantico. Emphasizing the importance of the team and its focus on the everyday warfighter, Colonel Irons, whose career had included commanding Marines at every level, stated in a meeting, "Nobody has a particular rice bowl [special project or group] that's more important than any other rice bowl. The only rice bowl that counts is Lance Corporal Binotz. By that I mean the guy on the ground. So we all need to work together."

This cultural creation of a fictitious faceless "everymarine" reflects the deep, underlying emphasis on the Marine as a member of a group rather than as an individual with a unique identity. Marines frequently greet each other by saying "Hello, Marine," an appellation that emphasizes their common identity rather than rank, name, or personal characteristics. Indeed, at least in theory, once one becomes a Marine, one's individual previous identity—one's ethnicity, race, gender, religion—no longer matters. As Lieutenant Colonel Gale, the commanding officer of a logistics battalion, affirms, "There are no demographic features to the Marines. We're all green [referring to the indistinguishable green uniforms Marines wear]."

Likewise, as one of the directors of the Marine Corps schools, Colonel Irons, states, "We are vanilla. We don't have any special flavors. All men are equal in the eyes of the Marine Corps. No one gets special treatment." He continued with an interesting story that illustrated not only the Marine Corps ideals of nonindividuality but also the cross-cultural misunderstandings that can result between U.S. services with different cultural ideals. "[A few years ago] the *Navy Times* came out with an issue. It had a cover titled the 'Magnificent 7' and featured the 'Magnificent 7' MEU [Marine Expeditionary Unit] commanders. Some guy was latching on to the movie which had just come out." Colonel Irons then leaned over his desk and paused to emphasize the disaster that resulted. "But when it came out you could have heard a pin drop. These things about individuality—I'm special. These aren't core traits."

The emphasis on lack of individuality is directly connected to the nature of combat. Early in my fieldwork and unaware of the extreme discomfort that bragging (or any semblance of it) would cause a Marine, I had asked Lieutenant Colonel Kramer a question about where he had received his many medals. We were at a promotion ceremony where he was wearing a formal dress uniform displaying the medals, and the ceremony had included a discussion about his role in the invasion of Iraq. Embarrassed, he began to speak and then stopped, mumbling that the medals didn't matter. He then turned the conversation quickly, noting that every Marine just does his job. In fact, he responded, no Marine can be unique or irreplaceable. For if he died in combat and he was the only one who could do that job, what would the unit do without him?

This self-concept of interchangeability is also connected intimately to the Marine Corps' values of adaptability and working together as a team. There is no job that a Marine is above doing, regardless of his or her training and skills. Major Neal, who was a foreign area officer (FAO), noted a bit ironically of his language and cultural expertise, "We focus more on creating generalists than specialists. If you make something a primary MOS [career specialty], then you lose the adaptability." "You have to be very flexible. Jack-of-all-trades. Master of all," Staff Sergeant Benson explained of the training he was leading for his unit's upcoming deployment to Iraq.

In fact, during my years working for the Corps I have observed many instances where Marines who have been trained with one set of skills have been placed in a billet (position) for which they had no training or preparation in order to fill a needed gap. Due to the changing and unpredictable nature of deployment, Marines are often required to step in and fill unexpected positions. One Marine who had an intelligence MOS (a job that studies and analyzes the enemy) described how he was assigned instead to the job of managing the budgets and payroll of the Iraqi army unit with which he was working. Another Marine who was a logistician (overseeing the logistical aspects of combat—food, water, transport, and so forth) became the mayor of an Iraqi town. A third Marine, who had an administrative MOS, was assigned duty in the hospital as a support to the medics. However, this flexibility in job tasking was just as evident back in the United States, especially after the government hiring freeze starting in 2010, which left many civilian positions vacant. During my years working for the Corps I have observed one Marine with a communications (radio, TV, computer specialist) MOS and another with a FAO specialization in the Middle East both get reassigned to manage the administrative sections and budgets of their units. A third Marine, who had an MOS as a tank operator, was given the job of supervising the renovation and construction of a new building.

"We use the term 'We do windows.' I've done too many things to pigeonhole me," stated Colonel Thompson, who certainly had lived a varied career traveling to well over fifty countries and conducting missions from combat to military training to disaster relief. He added, emphasizing his pride in this flexibility, "I wear the badge proudly of a general purpose force."

Because of this focus on creating a pool of competent generalists, there tends to be an avoidance of creating or retaining highly skilled specialists. "The emphasis on homogeneity means the Marine Corps doesn't tend to retain Marines with special skills. We have a limited capacity in civil affairs, engineering," explained retired Major Bates, now working as a contractor for Marine Corps Training Command. It also means that teaching and training are focused on the middle or average category of Marines rather than identifying, developing, and rewarding outstanding Marines with unique abilities.

"We always sacrifice talent to get an overall average playing field," Captain Nash noted as he discussed the training that his company was undergoing in 29 Palms, California. "The system hinders talent and challenges below-average bottom feeders. It works because we create units that can generally work. All are going to have the requisites to be a solid Marine."

Captain Nash's description of the Marine Corps may be somewhat harsh, especially since clearly the Marine Corps rewards talented and intelligent leaders, who are promoted into the senior officer ranks. However, his observations from his level—working with junior Marines—do reflect a reality that at the lower ranks, individuals who stick out, fail to conform to the norm, or challenge the system are likely not to be promoted or may even be expelled from the Marine Corps for insubordination.

This cultural pressure not to stand out is closely connected to Marines' extreme humbleness among one another and their frequent and colorful sayings suggesting that they are not very smart. A common way that Marines will present themselves to each other and outsiders is reflected in these comments by Colonel Irons: "We all know most Marines aren't real smart. We do a lot of grunting and waving hands." In fact, a popular term Marines use to describe themselves is "dumb grunt."

The more intelligent, perceptive, experienced, and senior Marines are especially likely to underplay their special talents or accomplishments. For example, Colonel Bedford, who had deployed numerous times to Iraq and

Afghanistan both as a commander of a battalion and in several joint positions, stated a very common (and obviously untrue) saying in a meeting of many senior officers when discussing a difficult planning problem: "It may just be that I am not the brightest bulb on the tree." Likewise, after a technical presentation on some communications software, Major Briggs, who was an absolute computer whiz, pointed to his partner and stated, "If he's a knuckle-dragging infantry grounder and I'm a slacker aviator and we can figure this out, you can see it is easy."

"Dumb grunt," "knuckle dragger" (both images referring to lower primate species such as gorillas), "low-watt lightbulb"—these are all terms Marines will use to present themselves to others. Not surprisingly then, overtly, Marines express an anti-intellectual culture. This does not mean that in reality Marines do not read, study, or think; in fact, some of my most outstanding students in my career as a professor have been in my classrooms at Marine Corps University. However, publicly one cannot indicate that one has been devoting a lot of time to reading or study. General Midway expressed frustration at this cultural bias when he stated, "If you went to a lawyer's office and he was reading a law journal you wouldn't think about it. If you went to a medical office and the doctor was reading a medical text you wouldn't think about it. But if I came into a military office and you were reading Patton we would say you were goofing off. It's a do, not think, culture."

And in this doing culture, the Marine who represents the ideal is not the smart analyst conducting intelligence[7] or the avionics whiz (although both are important members of the team), but the guy on the ground. And more specifically, the infantryman.

"It's All About the Guy on the Ground"

Although Marines move around the world by sea and are intimately connected to the Navy, culturally they view their role as fighting or resolving problems on land. While naval ships may take them to their destinations, their job really begins once they disembark. In fact, this distinction is not lost on the Navy, which has a translation for the acronym "MARINE"—"my ass rides in naval equipment" (suggesting that while on board ship, Marines are not particularly useful). Although Marines are proud of their ability to fight on land, at sea, or in the air, I have never heard anyone say, "It's all about the guy in the air," or "We're here to support the guys on the ships." Instead, my fieldnotes and interviews are laced throughout with comments such as "It's all about the guy on the ground"; "We're here to support the guy on the ground"; "The guy on the ground—he's the one that matters."[8]

There are varying interpretations of who "the guy on the ground" is depending on the discussion and context. In the more general sense, "the guy on the ground" is any Marine who is "forward deployed"—meaning living, traveling, and fighting outside of the United States. Marines make a clear distinction between those Marines who are working in safe jobs in the United States in "garrison" (that is, on their home bases) versus the Marine who is out there fighting the fight—being where the action is.

"Expeditionary culture—being in the fight, fighting. That seems to be the key part," explained Dr. White, a former Marine. Lieutenant Colonel Mason also emphasized this point: "My Marines like to fight. They're not really in this stuff for 'I love you man.' No they really want to fight." Captain Kraft, who had recently returned from Afghanistan, also noted the Marine desire to be where the action is, "That cultural association with the warrior class . . . looking for that combat, looking for that fight." Colonel Simons, who had traveled around the world in his twenty-nine years as a Marine, echoed a similar sentiment. Expressing the Marine identity of being expeditionary and out there fighting, he stated, "No Marine signed up to be a part of a garrison. We chose to name our main bases 'camps' [for example, Camp Pendleton, Camp LeJeune]. [It reflects] the idea that we set up and break down whenever we want."

This distinction (and even stigma) between those parts of the Marine Corps forces that are out there fighting versus those sitting safely in the United States was explained to me by Colonel Everett, who had just returned from deployment in Afghanistan: "There's a major division between the operating forces—those that do combat missions, warfighting—and the supporting establishment—those elements of the Marine Corps that create and sustain Marines. . . . There's a stigma to staying here [in the United States]. We use the term 'homesteading' [a term for Marines who stay too long in garrison]. But those Marines are necessary, though not necessarily important. The flip side is that the prestige, honor, sexiness of deployment is looked down on by the supporting establishment. 'You were out there as a commander having fun' [while we were stuck here taking care of you]. Of course it's not fun burying your Marines, getting calls at two in the morning."

In fact, being deployed and "out there fighting" is such a critical part of Marine identity that, according to Colonel Cole, who was commander of a weapons training battalion, "Marine generals do not want to be the next com-mandant of the Marine Corps. They want to be the combatant commander [the general who commands combat in a war overseas]. They want to be the warfighter." As a corollary, the importance of supporting the deployed oper-ating forces was described by Colonel Irons, whose job it was to oversee a Marine training program: "Anything that deals with the operating forces is a money maker. So strategy etc.—we have to do it. But we're 'boots on the ground' kinda guys. So we have to show what we're doing for the operating forces."

While the "guy on the ground" is always forward deployed, frequently, however, when Marines use this term they are referring specifically to the ground combat forces. These forces consist of four primary MOSs, each iden-tified by a specific two-digit number:[9] 03–infantry, 08–artillery, 13–engineers, and 18–track vehicles (for example, tank operators). Until very recently[10] women were not permitted to hold any of these MOSs (except engineer), and so in that sense the term "*guy* on the ground" is quite apt. In any deployment, the Corps will need Marines who have many skills to support these ground combat forces: finance officers and administrators to handle the unit's paper-work and finances; logisticians to help provide food, water, and shelter for the combat forces; intelligence analysts to figure out where the enemy is; com-munication specialists to manage the radio and electronic communications technology; and even public relations personnel to talk to the press. These, however, are combat support forces whose purpose is to support that guy out there actually undertaking the fighting.

The central role of the ground combat forces was brought home to me as my team began the statistical analysis of the Center for Advanced Operational Culture Learning (CAOCL) culture and language survey. At one point, one of the Marines working with me on the project pointed out that we needed to break our survey into two comparison groups—ground combat forces versus combat support forces. "You need to look at the narratives of the ground com-bat arms MOS—03, 08, 13, 18. In the Marine Corps these are the only MOSs that count. Tell me what the infantry guy says. Don't tell me what the avionics guy says."[11]

Interestingly, the example my colleague used for the "guy on the ground" was "*the infantry guy*," not the engineer or the tank operator or the artillery-man. This is not accidental, for in the Marine Corps there is a third meaning of "guys on the ground." When most Marines think of the quintessential "guy on the ground" that everyone is there to support, it's the guy with the rifle, and more specifically the infantryman. Colonel Thompson, who had just finished commanding a Joint Task Force, explains this ground and infantry focus: "I think we are 'enemy/mission centric.' Our predisposition to Marines on the ground relates to the fact that this is where man largely lives."

While all Marines are trained to use a rifle, not all Marines are infantry, a distinction that Colonel Simons made clearly in a discussion with me. "Rifleman does not equal infantryman. A rifleman is capable of defending him/herself and their unit in close combat, but he doesn't seek close combat—defense not offense. An infantryman [and ground combat unit] seeks close combat with an enemy—offense and defense."

Although, as mentioned earlier, no Marine is considered different or more special than any other Marine, the infantry is viewed as a highly valued and prestigious MOS in the Marine Corps. The high status of infantry, and particularly infantry officers, is indicated by the status that their school holds in the Marine Corps and among the U.S. services. "IOC, the Infantry Officers Course. It's the gold standard in infantry combat arms and leadership," Colonel Stacy, a retired infantry officer, stated.

One of the university faculty members, Dr. Green, explained to me on my very second day on the job, "The culture is very infantry centric. Every Marine a rifleman." In a separate interview, Colonel Everett noted, "The infantry subculture is extremely dominant. The commandant is always an infantryman. The Marine Corps is primarily an infantry organization." Indeed, throughout Marine Corps history, infantry officers have been the ones most likely to be promoted to general officer and especially to the position of the commander of the Marine Corps, the commandant.

However, while the infantryman may represent the "quintessential Marine" in public and Marine ideology, in reality only 18 percent of Marines actually hold this job in the Corps.[12] Not surprisingly, therefore, while many recruits may join the Marine Corps with dreams of engaging bravely in hand-to-hand combat, the majority of Marines will be assigned to hold another job or military occupational specialty. "In the Marine Corps we have more kids signing up to be infantrymen than we have slots for. They want to come in and 'do Marine stuff.' The good news is the aptitude battery [a screening test] works pretty well. The bad news is that we take the least performing Marines and slam them into the same MOS—truck drivers for example," stated Colonel Simons.

If 18 percent of the Marine Corps consists of infantry, then 82 percent of the Marines in the Corps must hold other jobs. Curiously, despite the number of highly specialized MOSs that require the use or maintenance of complex combat equipment, Marines continue to view the individual warfighter as far more important than any form of complex technology.

"Equip the Man vice Man the Equipment":[13] Technology, Specialization, and the MAGTF

> In the Marine Corps, the preference is for skills versus equipment. The caricature is that the Marine Corps would ideally go into battle with a spear and a loincloth. . . . And they don't even need the loincloth.
> —*Lieutenant Colonel (ret) Anson*

Lieutenant Colonel (ret) Anson's somewhat tongue-in-cheek comments reflect a Marine Corps ideal that places the individual human above the machine, that relies on Marine Corps savvy to solve the problem, not a computer. "It's not that we don't use technology when we need it," Colonel Simons, an infantry officer, explained to me. "But overreliance on technology over the human dimension is not something any Marine would want to do. . . . We believe in people first. That's why our Marines get the best training possible."

"The Marine Corps wouldn't exist without the human dimension," Major Neal, an Intel officer, explained to me. "We are human centric. It's the core essence of who we are. We have a very small budget so we can't be tool centric."

Although Marines realize that they must use contemporary technology to complete their mission, it is viewed as a means to an end. In fact, for some Marines, there is a bit of discomfort with the notion that technology could run decisions. "I don't want technology to drive me," stated Colonel Abrams in a discussion about reachback informational support for units deployed to Afghanistan. "I'm very suspicious of 'Being here and getting information from there.'"

"I don't want computer systems or processes to make decisions. I want people to make decisions," Colonel Irons stated during a meeting about training one day. "We've got to put a human in the loop. We run things."

"All of the technological innovations that were designed to help us with this war didn't amount to half of what the young Marines on the ground could do," Colonel Simons continued on with his earlier conversation. He then gave me an example of a new instrument that was supposed to show where everyone was on the battlefield. "It failed to provide information. In fact, it was misleading. Only one out of ten [of the instruments] worked. And you didn't see anyone as soon as you hit the palm grove."

This view that the individual thinking Marine is far more important than the technology he uses is reflected in a popular saying: "Marines don't man the equipment, they equip the man." The saying also reflects a subtle tension between the military's increasing need for improved technology and the potential for it to take over (as many sci-fi movies depict). "You're seeing the reascendance of soldiers over technology," Colonel (ret) Pierce noted in a planning exercise at MCU. "When you get down to it war is a people business. Magic bullet to the problem, magic arrow, magic sword—they don't exist. There is a tension love-hate relationship between the military and technology."

Yet today, while the physical toughness, perceptiveness, and courage of the individual rifleman is the commonly accepted ideal—both among Marines and in the general U.S. population—a second identity appears to be emerging. This new identity is being marketed to future Marines on current official Marine Corps websites and advertisements—that of the technologically superior Marine who bravely maneuvers sophisticated combat equipment to help in the fight. The U.S. government's Marine Corps Facebook[14] page does, of course, include the traditional fighting photos of Marines crawling through the jungle with blackened faces and wearing camouflage, Marines clearing a building in a military operations on urban terrain (MOUT) exercise, and Marines carrying and shooting rifles. Yet at least half of the initial fifty photos that I examined also emphasized the use of "cool technology." The picture gallery includes various helicopters and fighter jets whizzing through the air, Marines shooting off a M777 Howitzer and other artillery, and Marines driving in MRAP (Mine Resistant Ambush Protected) vehicles. Recent Marine Corps video/TV advertisements reinforce these visions of a high-tech force.[15] These catchy, often fast-paced videos flip rapidly through action images—not only of Marines scaling walls and shooting rifles but also of helicopters swooping down from the sky, amphibious assault vehicles landing on beaches, tanks rolling through deserts, Marines scoping out the enemy with night vision goggles, and fighter jets taking off from the decks of aircraft carriers.

This role of the tech-savvy Marine much more realistically reflects the contemporary reality for the majority of Marines in the Corps. According to recent statistics by the Marine Corps, over 47 percent of Marines hold jobs that require the use or maintenance of complex combat machinery, technical equipment, and vehicles, including such occupational fields as communications, artillery, engineer, tank, motor transport, signals, ground electronics, electronic maintenance, aircraft maintenance, avionics maintenance, airfield services, air control, navigation, and pilot.[16]

Interestingly, and perhaps suggesting a slight shift in Marine Corps ideals to reflect the valuable role of a technologically capable Marine, the 35th Commandant of the Marine Corps, General Amos, is not an infantryman, but a pilot. Perhaps this is not surprising, considering that 17 percent of all Marines hold MOSs (ranging from maintenance to pilot) in the air wing (plane and helicopter) component of the Marine Corps—almost exactly the same percentage as the infantry.[17] And pilots, like the ground combat forces, do engage in direct combat with the enemy. However, during the Commandant's tenure I have occasionally heard a few muted comments about the *aviator* commandant, suggesting a somewhat ambivalent attitude among some Marines toward a noninfantry leader. This tension, in fact, indicates a subtle rivalry between the air and ground combat forces as reflected in the numerous jokes made on both sides about the other (for example, Major Briggs's comments about the slacker aviator and the knuckle-dragging infantryman). Perhaps this is in part because pilots appear to view the battlespace and the role of technology in it somewhat differently. As Major Neal points out, "Pilots don't want to be tied down to the AO [area of operations], but if you ask the majority of the guys on the ground they still prefer the low tech."

Despite this subtle internal rivalry, Marines are quick to point out that ultimately what binds the Marine Corps together, indeed one of the aspects of the Corps that makes it unique among the services, is their ability to work as a team, combining the individual skills of the various military elements—ground, air, and the supporting logistics—into one smoothly functioning combat task force called the Marine Air-Ground Task Force (MAGTF).

"No Marine Left Behind": The Marine Corps Family, Teamwork, and the MAGTF

Being a Marine is all about being a member of a team, about being a member of a close group that cares about you and is willing to protect you and die for you. Taking care of one's fellow Marines is a bedrock component of Marine Corps ideology. "No Marine left behind" is not a slogan to improve Marines' education but a promise to all Marines that in combat and back home, the wounded and dead will not be abandoned or forgotten. A recent Marine poster displayed in the halls of MCU shows a picture of a fire team of four holding a wounded comrade in their arms. The subtitle reads: "What have you done for him today?"

"You need to show your Marines you care about them," Staff Sergeant Duke told me as we discussed training for his platoon preparing for deployment to Iraq. "No one wants to fight with someone you don't care about."

The importance of developing bonds of caring and commitment among Marines is so great that these bonds are often described in terms of family, especially among enlisted Marines. Marines refer to themselves and their relationships with their junior Marines using a variety of kinship terms, as the following conversation among three infantry staff sergeants illustrates:

STAFF SERGEANT DUKE: You've got to do what you can to help them out to a reasonable extent and create that trust. You've got to be like a hard hard father. You can't be nice to them all the time, but . . .

STAFF SERGEANT KUKELA: You have to let them know that you care about them.

STAFF SERGEANT BENSON: Yup. It's a toss-up then. You have to tell them what to do, have to be strict. But they got to know you care about them because if they don't think you care—

STAFF SERGEANT KUKELA: We lose the fight if someone don't care about them.

STAFF SERGEANT DUKE: And sometimes you have to be the bad guy, but they usually understand that if you're not just out there looking to screw with them or mess with them.

STAFF SERGEANT KUKELA: Yeah belligerent or not you're their father.

STAFF SERGEANT BENSON: Mother, brother, sister too.

STAFF SERGEANT DUKE: You're their father, their counselor, everything. . . .

STAFF SERGEANT KUKELA: You've been dealing with them so long it's almost like they're your family. They're no longer your platoon. This is my kin, my brother.

The slogan "No Marine left behind" not only emphasizes the deeply felt, almost familial commitment that Marines feel for each other but also underlies the importance of loyalty to a larger purpose: the Corps and their country. "What distinguishes the military from civilians?" Captain Johnson, a student at the Expeditionary Warfare School, posited. "A belief in something greater than yourself. A feeling of belonging—that it's something important, really important, you're contributing." A second student, Captain Nelson, added, "When we [Marines] say loyalty, it's not just being loyal to a company or team. But also being loyal to the Corps." In fact, a popular way for Marines to prioritize their values and loyalty is "F2C2"—faith, family, country, Corps.

The important concept of self-sacrifice and contribution to the greater good was illustrated by Colonel Irons's comments when talking to me about the challenges of teaching young recruits the values of the Corps. "We didn't promise you a rose garden. We're not here for you [the individual Marine]. You're here for us. It's never about the individual. It's about the team. . . . I'd rather have a team player who gets along with everybody than someone who is a star and thinks he's hot stuff and can't work with anybody."

This concept that a team player is essential to the successful functioning of the unit was brought home during a MCU guest lecture on senior leadership. General Taylor was giving advice to the students on managing a joint staff composed of members from various military and civilian services: "What if you've got an actor from G-3[18] [operations division] or whatever who's not a team player? Just wants to do the job and go home? Get rid of him. Don't let them send anybody up that won't roll up their sleeves and get to work."

A popular Marine Corps saying, often repeated when a group is not working well together, is, "There is no 'I' in team"—in other words, there is no place for the individual on the team. Likewise there is the expression "Everyone reading off the same sheet of music." "It means everyone is working together, creating a harmony," explained Dr. Green, a member of the MCU faculty.

Teamwork is so critical to the Marine Corps that developing team players is a major focus of initial training at both the recruit depots (see chapter 4) and Officer Candidates School (OCS, see chapter 5). At OCS, teamwork is one of the Five Pillars in training. In fact, other candidates' opinions of their teammates are a key factor in evaluating potential Marine officers. "The platoon staff evaluates what the other candidates think of the candidate. It's not just the platoon staff's decision. The focal view is how other people see it," explained Master Gunnery Sergeant Puller, an OCS drill instructor.

At the recruit depots, learning to be a team player is an important part of the drill instructor's job. "Phase One is about taking the individual identity out of them and [learning to] lean on the team," Colonel Simons, a former commanding officer at the Marine Corps recruit depot at Parris Island, explained. "Every recruit has that point in recruit training. . . . He real-izes 'It's better if I help my buddies.' For a lot of them it happens about six weeks. . . . You can tell when a unit starts to act like a team. The DI [drill instructor] starts to shift focus. He looks for that break: [the moment] when they see that 'It's easier when we work together.'" He continues, "The team is paramount. Why? In the end no individual Marine can accomplish the mis-sion. Only the team can accomplish the mission."

The importance of the team is not limited to individual Marine units, however. In fact, teamwork is the foundation of one of the most important successful innovations of the Marine Corps way of fighting—the MAGTF. As I discuss in the following chapter, the MAGTF is a unique integrated team consisting of three elements that fight together in a carefully orchestrated plan to combat the enemy: the ground forces, the air component (helicopter, airplanes, and other air support), and logistics (the practical support supply-ing fuel, water, food, and equipment).

"Marines created the MAGTF where all of the units—air, ground, logis-tics—work together as one team. The importance of the MAGTF is that you fight as a holistic group. You're totally integrated," Colonel Simons stated in an interview. Then he added a most intriguing comment, "The strength is that when the guy on the ground talks to the guy in the air—he's got a vested interest in me. Because he *is* me. He's a Marine."

This concept that the individual Marine is, at heart, the "everymarine"—connected by indestructible bonds that transcend time and place—provides great insight into the meaning of the famed Marine Corps sayings, "Once a Marine, always a Marine" and "You can take the Marine out of the Corps, but you can't take the Corps out of the Marine."

Having grown up in an extremely individualistic American culture, the notion that in the Marine Corps one's identity could at some level be merged and indistinguishable from one's teammates was most puzzling to me. Colonel Thompson shed some light on his way of seeing the world when he stated, "We may come from the cross section of society but we refuse to reflect it. . . . From our ethos, the heart of the Marine is that individual rifle-man. That's why we have TBS [The Basic School for all officers] so that every Marine learns to use a rifle and undergoes the same experiences. Then when you go on to be a pilot, for example, and you see Marines on the ground, there's an understanding of that situation. He's carrying a fifty-pound pack, [and you know from your own experience] that if the terrain is uneven he will twist an ankle—so he has to move slowly [and so do you]. That's why there is no cultural difference between the guy in the air and the guy on the ground." He then concluded by emphasizing again the critical importance of the individual "everymarine"—the grunt, the guy on the ground, the rifleman: "Lance Corporal Binotz transcends the MAGTF. The MAGTF is one team with all these capabilities. But at the end of the day if you want to hold ground, it's got to be the guys standing there."

3 "Soldiers of the Sea"

The Marine Corps Expeditionary Mindset

From a post on the 26th MEU webpage, September 15, 2010:[1]

USS KEARSARGE, AT SEA —"Reveille, reveille, reveille," a high-pitch, garbled voice loudly announces over the intercom, sending reverberations throughout the compact living area. Six a.m. and Lance Corporal Nicholas A. Taft is awake, though his body rebels at the adjustment as the ship sailed east through another time zone the night before.

The Marines and sailors with 26th Marine Expeditionary Unit (MEU) have trained together for five months preparing for their current mission. Taft, a landing support specialist with Combat Logistics Battalion 26, 26th MEU, climbs out of his individual sleeping space: a Navy "rack" half the width of most children's beds and stacked three or four high against another row of racks.

"Everything is pretty tight living in berthing," said Taft. "If the people around you weren't your friends before, they are now."

Trying not to hit the steel outline of his sleeping area or any of the other hard surfaces in the small space, Taft maneuvers out of his rack and into the row. He shares his living space with more than 100 of his fellow Marines. Each of them tries to maintain his footing as they work around each other going to the "head," the Navy's version of a restroom, to brush their teeth, shower, shave and put on their uniforms as the deck beneath them constantly rocks as the ship sails the ocean.

Marines and sailors move about the narrow passageways and try to avoid slamming their shins on the "knee-knockers," raised steel barriers at hatches separating sections of the ship. They ascend and descend steep ladderwells holding onto the handrails, anticipating the slow rolls the ship might make.

After a 45-minute wait in line for morning chow (meal), they proceed to their work spaces, squeezing past each other in the passageways and sending out a greeting of, "good morning," or, "Oorah!"

As the ships of Kearsarge Amphibious Ready Group make their way across the Atlantic Ocean to support humanitarian assistance efforts in Pakistan, 26th MEU Marines continue to live, work and conduct training and operations aboard the ships.

FROM AUGUST 2010 TO MAY 2011, WHILE MOST OF THE world's attention was focused on the wars in Afghanistan and Iraq, the 26th MEU sailed to four continents and conducted missions ranging from disaster relief to foreign military training to direct combat. In its nine months of deployment, the MEU not only held scheduled training exercises with partner militaries in Kenya, Jordan, and the United Arab Emirates, but it also provided emergency humanitarian assistance, delivering food and supplies in Pakistan in response to one of the "worst flooding disasters in history";[2] deployed Battalion Landing Team 3/8 to Afghanistan to augment the U.S. military surge in Helmand Province;[3] and supported Operation Odyssey Dawn—protecting civilians during the civil war in Libya and rescuing a downed U.S. Air Force pilot.[4]

As the maritime title—marine—implies, the Marine Corps has always been intimately connected to the sea and its partner service, the U.S. Navy. In recent years, Marines have become perhaps best known for their roles in the wars in Afghanistan and Iraq. However, as the two most recent commandants of the Marine Corps have stated,[5] the primary purpose of the Corps is to be ready and able to act as a sea-based emergency response team—the first to be there in a crisis, whether on land, at sea, or in the air. "We are by nature amphibious," a faculty advisor at Marine Corps University , Lieutenant Colonel Hall, explained to me. "While we've been in Iraq and Afghanistan lately that's not what it's about."

"We're amphibious. We're littoral. We're not big Army," noted Colonel Irons, who had spent much of his twenty-six years in the Marine Corps traveling around the world on various missions.

Marine Corps literature describes the military organization as "expeditionary"—a term it uses to describe the Marine Corps' approach to conducting rapid, lightweight "expeditions" to dangerous foreign places.[6] "Marines have always had an expeditionary mindset," said Colonel Simons, who is currently assigned to the Marine Corps headquarters in Washington, D.C. "We didn't go west [like the Army]. We always went elsewhere—the Barbary Coast, the Mexican-American War, the Banana Wars."

Today, as in the past, Marines still travel everywhere, ready to help resolve (or at least mitigate) the world's problems. As this chapter illustrates, this expeditionary and amphibious purpose is intimately linked to the Marine identity of always being at the forefront—ready, agile, flexible, and able to handle any crisis. A second related identity—that of the hard, lean Spartan—can also be seen as a corollary aspect of Marines' self-image as expeditionary leaders in a small force that moves lightly and rapidly around the world to achieve the mission without the burden of heavy creature comforts. These two identities—tough, lean Spartans and agile leaders—are, in turn, reproduced and reinforced through a unique set of flexible organizational structures, which morph according to the changing needs of Marines' expeditions and operations around the world.

Many of the studies of organizational structures (whether military or civilian) focus on the way that organizational *culture* is influenced by organizational *structure*.[7] In this logic, individuals' behavior within the organization—including responses of individuals and groups within the organization to policies implemented from above or outside—is directly influenced and constrained by the structure of the organization. This assumption has led to the almost epidemic reorganization and restructuring of many governmental organizations and corporations in recent decades.

As the case of the Marine Corps illustrates, however, organizational culture can also determine and influence structure. In the Marine Corps, the focus on mission accomplishment supersedes everything. As a result, Marine cultural ideals of adaptability, austerity, and sacrifice in order to meet the mission result in innovative and ever-changing structures to achieve the task. Structure, then, is a secondary by-product of Marine Corps culture and the Marines' identity as the "first to fight."

"First to Fight": The World's "911 Force"

On November 10, 1775, so the story goes, the Second Continental Congress met in Tun Tavern in Philadelphia and commissioned Samuel Nicholas to establish two battalions of Marines. These Marines would act as naval infantry ("soldiers of the sea"), maintaining order on ships and disembarking to fight wars on land. Legends, of course, tell us more about the ideals and identity of a culture than perhaps cold historical facts. And indeed evidence suggests that Marine Corps recruiting actually began somewhat earlier at Conestogoe Waggon on Chestnut Street.[8] But the image of a swashbuckling, hard-fighting, and hard-drinking group of men who signed up in a tavern for adventures at sea and in faraway lands is far more romantic and suited to Marines' views of themselves. Indeed, this image persists today, as recent comments in a lecture by General Spaulding reveal: "Marines talk, swagger and most importantly, fight today in the same way America's Marines have since Tun Tavern."

Although the Marine Corps is separate from the U.S. Navy, with its own budget and commander (the commandant), as the description of the 26th MEU's voyage to Pakistan highlights, today, as in the beginning, Marines and sailors frequently partner together in order to carry out the Marine Corps' expeditionary tasks. The Marine Corps' strong naval ties can be seen both in its colorful naval language and in its requirement for all Marine recruits to pass a challenging swimming qualification (which includes jumping off a ten-foot diving board with hands and arms tied and "swimming" back to shore). In daily speech, like their seafaring partners, Marines use naval terminology to refer to their physical surroundings. This applies even to Marine buildings on land. As I quickly learned on my first day on the job, floors in Marine Corps buildings are called "decks," doors are called "hatches," stairs are "ladders," and bathrooms are "heads."

Interestingly, while the two services each have their own separate budgets, they do share some personnel, creating an economic as well as a cultural and historic bond between the services. The U.S. Navy provides all of the medical services (including doctors and Corpsmen), some of the engineers, and all of the chaplains for the Marine Corps. As a result, most Marine units usually have naval personnel such as chaplains and doctors embedded with them. This does not mean that Marines and sailors are interchangeable, however. With the exception of loaning out skilled personnel, the Navy's sailors run ships and usually fight their battles at sea, while Marines typically are prepared to conduct most of their operations on land—although based from the sea.

It is this concept of basing Marine operations from the sea that connects the Marine Corps with the Navy. The Navy not only carries Marines and their combat equipment to destinations around the world, but the two services frequently deploy together for six-to-twelve-month periods in a special configuration called a MEU. These MEUs can be viewed as sailing crisis teams that float on regular routes around the world, ready to deploy Marines immediately to any destination where they are needed. As the 26th MEU's voyage illustrates, the MEU's circuit usually includes a certain number of scheduled stops. For example, in the case of the 26th MEU's 2010–2011 tour, Marines conducted scheduled joint training exercises with the militaries of Kenya, Jordan, and the United Arab Emirates. However, during the voyages of the 26th MEU on the Navy ships of the Kearsage Ready Group, the Marines also responded to several unplanned crises: delivering humanitarian aid in Pakistan, supporting combat operations in Afghanistan, and conducting joint peacekeeping operations during the Libyan civil war.

Reflecting this emergency response identity, one title used for the Marine Corps in its documents and media materials is "America's force in readiness." In fact, the Marine Corps is chartered to be the force that is "the most ready when the nation is least ready."[9] Dr. White, one of the professors at MCU, emphasized this emergency response identity: "We need to be the 911 force. We need to be ready to go somewhere [at a moment's notice]. Whenever there is a need to be somewhere . . . the Marines are the amphibious force waiting offshore."

"[It's our] expeditionary ethos," explained Lieutenant Colonel Anson, who had retired from the Marine Corps and was now working in a Washington, D.C., think tank. "You need to be ready physically and mentally for where you're going to go tomorrow. You're not going to get clear direction. You're not going to hear where you're going [until you are there]. You're going to have to go where you have to be."

This need to be instantly prepared at a moment's notice underlies a central Marine value: combat readiness. An exhibit on "Life aboard ship" at the National Museum of the Marine Corps in Quantico, Virginia, holds the following caption: "Combat readiness. This tradition of rapid deployment of ready forces in a national emergency has become a hallmark of the Marines."

In fact, one aspect of the Marine Corps that is often unseen or unnoticed by many Americans outside the Corps is the degree to which Marines rapidly respond to an enormous variety of crises ranging from fierce combat to humanitarian aid. In the past decade alone Marines have not only fought in two major combat operations in the Middle East (Iraq and Afghanistan) but also have been busy conducting peacekeeping and humanitarian operations, military training exercises, noncombatant evacuation operations (NEOs), stability operations, civil affairs programs (engineering and aid projects such as building roads and inoculating cows), and many other projects around the world. In the first decade of the twenty-first century, for example, while engaging in combat operations in Iraq and Afghanistan, Marines also responded to an earthquake and tsunami in Japan (2011) and Indonesia (2008); undertook the immediate evacuation of Americans during an uprising in Lebanon (2006—as well as similar operations in 1976 and 1982); engaged in a combined North Atlantic Treaty Organization (NATO) peacekeeping operation to protect civilians in Libya (2011); and even made international headlines for their rescue of a civilian ship taken hostage by Somali pirates (2010). In all of these cases, Marines were required to respond almost instantaneously to an immense variety of world crises—often in less than twenty-four hours.

In the 2010 Center for Advanced Operational Culture Learning (CAOCL) culture and language survey, we asked Marines to list the places they had deployed. We also asked them to identify the kinds of missions in which they had participated. Figures 3.1 and 3.2 illustrate their responses, reflecting the extremely varied work Marines undertake all over the world.

According to a study of Marines' humanitarian/peacekeeping missions, the Corps conducted, on average, at least one such operation per year between 1900 and 1994.[10] Interestingly, however, until recently this more humanitarian aspect of Marine operations has generally been neglected in Marine advertising and remembered history. In fact, a stroll through the National Museum of the Marine Corps, which can be viewed as a repository of the most important moments in Marine Corps history, reveals not a single exhibit on a Marine Corps humanitarian or peacekeeping operation. Virtually every exhibit focuses on the major battles of the Corps, suggesting that the primary identity of the Marines resides in their fierce combat prowess in war, rather than their more diverse emergency response, partnering, and peacekeeping missions.

While Marines are typically the first to arrive in a crisis, theoretically, Marines are not there to create a permanent solution or to definitively resolve the situation. Instead, according to a U.S. government mandate,[11] Marines are intended to be a small advance force that deals with the immediate crisis,

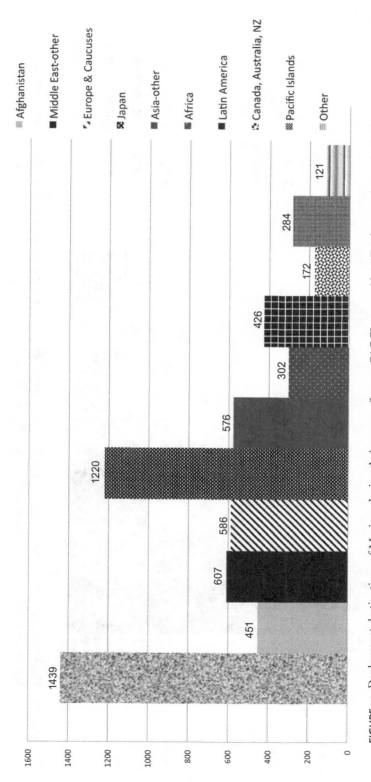

FIGURE 3.1 Deployment destinations of Marines during their careers. Source: CAOCL survey. Note: Total respondents who had ever deployed were 2069. Marines could list more than one region resulting in a higher total N.

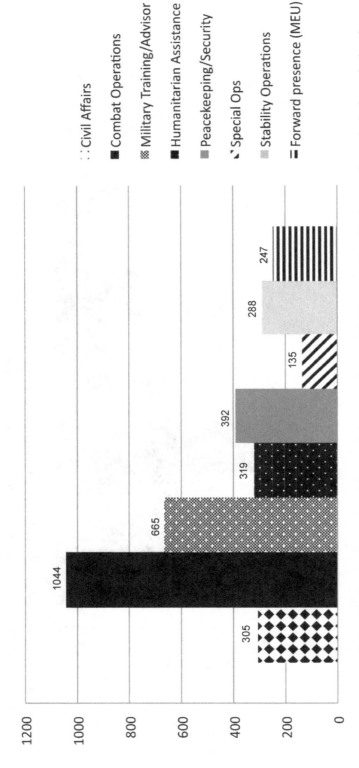

FIGURE 3.2 Missions Marines have participated in over their careers. Source: CAOCL survey. Note: Total respondents who had ever deployed were 2069. Marines could list more than one mission resulting in a higher total N.

leaving the longer-term solutions to the other services or the State Department and non-governmental organizations (NGOs) that follow behind.

Colonel Simons describes the difference between the expeditionary role of the Marine Corps with its "light footprint" and the role of the Army, which is expected to move in and finish the job. "This is the whole notion of expeditionary. . . . The Army, if they go, they go and stay. For us [the Marine Corps], there are not going to be many of us, and there will be a lot of you [the foreign population]. . . . The Army is the guarantor of your military. They are the one that won World War II. They won the West. They built the railroads." In metaphoric terms, then, the Marine Corps serves as the emergency room medical team of the world hospital: conducting triage and then, ideally, handing the patient to other organizations with more resources to resolve the problem. Increasingly, however, as the wars in Afghanistan and Iraq have illustrated, today's Marines are not only expected to be the first to arrive in a crisis but also to remain and assist their joint partners in resolving the problem.

According to the Marine Corps publication "Marine Corps Vision and Strategy 2025,"[12] this kind of varied work in "complex hybrid conflict environments" is expected to be the norm for the coming decades. The Marine Corps anticipates that throughout the twenty-first century, it will continue to respond to small, episodic conflicts and crises requiring "stability and support operations; small wars and counterinsurgency; humanitarian assistance; peace operations; counter-proliferation (of nuclear weapons); combating terrorism, drug trafficking and crime; and noncombatant evacuation operations."[13]

Given the immense variety of operations that Marines must conduct anyplace and anytime, it is not surprising that a central Marine ideal is to be agile and flexible. In an interview with Major Bennett, who had served on a military transition team in Iraq, he mentioned, "As a Marine it's important to be creative, be flexible, be resourceful." And Captain Hutchinson, who had been part of the initial invasion in Iraq, stated, "A Marine is adaptive, someone who can respond to a myriad of situations." These ideals were reiterated by Major Perkins, an instructor at one of the Marine Corps schools, who noted, "Adaptability. Our culture is all about adapting, improvising,"

The Marine ideal of being flexible and adaptive is so ingrained that there is even a specific Marine term for this concept. Playing on the Marine saying "Semper Fidelis" (always faithful—see chapter 4), Marines have coined a new term referring to the 1970s bendable stick figure, Gumby. "[There is] a willingness to deal with ambiguity, uncertainty. You might be mayor of a town one day, and give birth to babies the next," stated Lieutenant Colonel (ret) Anson. "I think 'semper gumby' sums up that. We take tremendous pride in our expeditionary nature. It's a central part of who we are and what we have become."

Interestingly, flexibility and adaptability are seen as based on and deriving from discipline and commitment to a shared purpose. As Colonel Thomas, a commanding officer at one of the training programs, observed, "The corollary is that to be in combat you have to be adaptable. You combine that with a level of self-discipline . . . the installation of discipline and physical toughness to be adaptive in combat."

Paradoxically, then, discipline creates the ability to adapt—*the opposite of* rigidity and inflexibility. Captain Rhodes, speaking about his experiences in Iraq, challenged stereotypes of a rigid military when talking about the Corps, particularly in the complicated combat situations Marines face today. "Having been on a police transition team, I can say that rigidity of schedule, the lockstep military dynamic does not apply at all. In our case, we threw a dart out there, and now we're coming back to adaptability." "In the Marine Corps we have an inherent flexibility that the Army doesn't have," Colonel Simons noted. "The Army has to be the winning force. At the end of the day, the Marine Corps is not going to save the country."

General Nassau elaborated on this concept by explaining that Marines all share the same commitment to the goal or endstate; however, their flexibility makes it more possible to reach the goal successfully. "We adapt to the situation. The way I see it, we adapt to accomplish the task we have. We take the endstate and work it out to reach it. Maybe not so much back here [in the United States], but out in combat, we get results. We won't necessarily follow somebody's rules, but we get results."

Contrasting the Marine Corps' expeditionary nature to the more permanent stationing of the Army and the Air Force, Colonel Simons described the practical issues that the larger and less mobile U.S. services face when they deploy. "This whole notion of expeditionary. . . . The Air Force came up with the AEW—an air expeditionary wing—but how are you going to do that? They still need a 10,000 foot runway. They need hundreds of gallons of fuel. They need their air-conditioned buildings. The reality is that they are not expeditionary."

Noting, however, that even the Marine Corps is heavier and less mobile than it would like to be, Colonel Simons added, "We're running convoy after convoy of fuel up to Afghanistan [to support high energy use bases]. But it's costing us $2 to ship $1 of fuel up there. And we're getting blown up [on the way]; and we need this huge project to get it there. . . . [So] we're looking for new ways to set up small, low energy bases. Better solar panels, better lighting, better insulation for tents."

Over the past few years the Marine Corps has been conducting a number of research projects to reduce its "footprint" (its effect on and use of the environment) under the acronym EX-FOB. The acronym is quite appropriate given the Marines' identity as a mobile and light force that operates from ships and tents, rather than residing on a semipermanent basis on forward operating bases (FOBs). Not without reason, Marines speak of the Corps as the "lean green fighting machine"—although historically, "green" has referred to their clothing and not to their environmentally sensitive policies. A recent rather hilarious poster produced in collaboration with the U.S. Fish and Wildlife Service (see Figure 3.3), however, suggests that perhaps a new, more ecologically conscious Marine identity is emerging today: one that acknowledges that to be lean and mean, they must also be "green."

"The Lean Green Fighting Machine"

> Marines are aggressive, austere. It has to do with what I see as the Marine Corps' purpose as first to be there. We don't expect to bring a lot of comforts or amenities with us.
> —*Captain Hutchinson*

The Marines' identity as an emergency expeditionary force is not only manifested in their ideals of flexibility and adaptability but also in their ideals of leading a lean, hard Spartan lifestyle. In order to be agile and move quickly, Marines believe that they need to be prepared to live without the comforts of modern life.

This Spartan ideal of tough, hard men and women who are inured to pain was expressed in numerous ways in my interviews, most particularly in discussing the physical difficulties of deployment. "That's the reason Marines only deploy for seven months," explained Colonel Monroe, a commanding officer who had deployed several times to Afghanistan. "We believe we can go out there, live off the local economy, go without showers, live hard for seven months. We can drive our boys hard. They can live dirty, sleep in a hole, not get the Internet for about seven months. Then you've got to send them home."

FIGURE 3.3 Lean Green Fighting Machine. Source: US Marine Corps and US Fish and Wildlife Service

During my years as a professor at MCU, I have heard numerous debates about which of the Marine Corps bases is the toughest and hardest (in the process reinforcing the ideals of suffering and endurance). In one such discussion, Colonel Everett argued, "29 Palms is the real Marine Corps. First, I admit I'm jaded because I'm a 29 Palms guy. I've done two tours out there, commanded out there. The quality of training is better than anywhere else— infantry training. Plus everything else is harder there too. The environment is hard. It inures young Marines to some of the habitual skills they need to have. Keep hydrated. The desert will take you in a minute if you're not careful. It's unforgiving."

Colonel Thomas described the importance of training Marines through hardship to prepare them for combat: "Warrior toughness and resiliency. You only get resiliency through repeated hardship. You've gone through difficult things in your training [and you can handle it]. The more you sweat in peace the less you bleed in war." Similarly, Colonel Everett, another school director, observed, "The reason we try to make training as hard as possible is that we can never replicate how hard it will be in the field. We try to inoculate Marines against violence a little bit, but you still cannot replicate the stresses in combat."

"Another saying is 'through pain there is discipline,'" Staff Sergeant Hanson, a drill instructor (DI) at Parris Island Recruit Depot, observed in discussing the role of DIs in creating Marines from civilians. "When one recruit messes up, everyone pays. If one recruit takes crackers from the chow hall, then everyone gets punished. So after a while they start taking that moral responsibility. It really is a lifestyle we live." Emphasizing the aspect of developing Marines through hardship, a movie on recruit training at the depots described recruit training as "a twelve-week battle between pain and persistence."[14]

Being hard and pushing Marines to develop strength and resilience is seen as a positive action showing that leaders and instructors care about their Marines. "Marine Corps tough love. Frowning and being stern doesn't mean you hate them," explained Captain Bailey, who was discussing training of officers at Officer Candidates School. Suffering together is viewed as an important way to form bonds of brotherhood between both leaders and their junior Marines. In a discussion about leadership, Captain Edison, a student at Expeditionary Warfare School, noted that an important characteristic of a leader was someone who was willing to share the same suffering moments.

"[Good leaders are] guys who just share hardships with them [their Marines]. Lead by example. You don't always have to lead, you just have to be there."

In speaking about their ethos of toughness and endurance, Marines that I interviewed often likened themselves to the Spartans, whom they saw as sharing the same characteristics of loyalty and honor hewn through shared suffering and hardship. "We are always faithful to each other—arm in arm like the Spartans," noted Lieutenant Colonel Mason during a discussion about Marine Corps culture with several foreign military officers.

This notion of an inseparable bond, forged through hardship, was reflected in Marines' willingness and desire to live and work under austere conditions even when not deployed. In a manner similar to sympathy hunger strikes for fallen or jailed comrades, the Marines I worked with voluntarily sought to go without the luxuries of American life on base in the United States to demonstrate their support for their fellow Marines who were deployed overseas. For example, given the rapid increase in both civilians and Marines working for the Corps in recent years, both civilians and Marines on the base in Quantico, Virginia, often found themselves working in extremely cramped and uncomfortable conditions—sharing small, windowless offices; working in old, leaking, poorly heated or air-conditioned portable trailers; and lacking desks or computers or other essential work materials. Although some of the civilians complained occasionally, I have no recorded instances where any of the Marines I worked with openly expressed discomfort or frustration with the situation. To the contrary, they would often remind me or other civilians (who expressed irritation at the latest failure in Internet connectivity or a new leak in the roof) that this was luxury compared to the combat conditions in Afghanistan or Iraq. "Suck it up" and "Figure it out Marine" were common responses I would hear whenever a Marine would even slightly indicate dissatisfaction with the situation.

The comparison between the Marines and the Spartans is not only based on their shared ideals of a tough, hard, austere fighting ethos but also on their view of themselves as a lean, small, elite fighting force always under threat. While numbers vary from year to year, the Marine Corps is the smallest of the four primary military services (excluding the Coast Guard and several focused military programs) and receives the smallest budget relative to the other services (see Table 3.1).

As General Krulak describes in his book *First to Fight*,[15] the ability of the Marine Corps to accomplish what it does, despite its relatively small size, is a

TABLE 3.1 Budgets and personnel of military services compared

Entity	Budget 2012[1]	Percentage of budget	Total military personnel 2013[2]	Percentage of personnel
Army	$203.2 billion	38%	552,100	39%
Navy	$142.0 billion	26%	322,700	24%
Marine Corps[3]	$31.6 billion	6%	197,300	14%
Air Force	$163.1 billion	30%	329,460	23%
TOTAL	$539.9 billion	100%	1,401,560	100%

SOURCES: Budget: Office of the Under Secretary of Defense "National Defense Budget Estimates for FY 2014 (Comptroller), May 2013, 24, 28, and 32. http://comptroller.defense.gov/defbudget/fy2014/FY14_Green_Book.pdf. Personnel: Office of the Under Secretary of Defense "National Defense Budget Estimates for FY 2014" (Comptroller), May 2013, 53. http://comptroller.defense. gov/defbudget/fy2014/FY14_Green_Book.pdf. Marine Corps budget: Headquarters Marine Corps Programs and Requirements, personal communication.

NOTES:
1. National Defense Budget estimates do not break down the navy budget between the Marine Corps and Navy. The total navy budget allotted is $173.6 billion.
2. Figures do not include reservists, civilian personnel, or contractors working for each service.
3. Marine Corps figures do not include what the navy provides to the Marine Corps in terms of general support (ships, some manpower, some aviation) referred to as "Blue (Navy) on Green (Marine Corps)" support.

source of pride and a central part of the Corps' identity. Colonel Thompson, who had recently finished a joint assignment working with military members from a number of services, pointed out that Marines are "efficient and effective, [providing a good] bang for DoD [Department of Defense] dollars." Indeed, as Table 3.1 illustrates, Marines provide approximately 14 percent of the U.S. military manpower but only cost the government around 6 percent[16] of the U.S. military budgets.[17]

Rather than viewing their relative lack of money and small size as a detriment, the Marines in my study viewed their austerity as a positive aspect of life in the Corps. "Being thin is good. Being thin forces you to prioritize," noted Colonel Chase in a talk about the challenges of operating in Afghanistan. Major Holbrooke, who had deployed numerous times as an engineer on various small Marine Corps landing teams, noted, "Making us smaller makes us a tighter group."

However, as Krulak[18] describes in great detail, the Marine Corps' small size and versatile purpose have, at several points in history, almost been its downfall. In contrast to its sister services, the Marine Corps has no obviously unique combat capability. Both Marines and the Army fight on land; the Air Force, Navy, and Marines all have their own squadrons of combat and transport planes and helicopters; and while Marines do sail around the world to conduct their business, the Navy owns and runs the ships and fights naval battles, not the Marines. Consequently, during various points in U.S. history, certain members of Congress have proposed that the Marine Corps is redundant and an unnecessary cost. The result, according to Colonel Simons, has been an ongoing "concern for the organizational survival of the Marine Corps. The Marine Corps is small and has a sense of extermination."

Major Neal, who was working in a joint position in Washington, D.C., explained that "an organizational paranoia animates both the Marine Corps and the Coast Guard. There were ongoing attempts to abolish the Marine Corps up until World War II." True to the Marine Corps "can-do" ethos in which hardship and struggle are good, he then added, "The competition . . . is probably healthy for us. It feels like we are under threat. It forces us to deal with [new situations], to adapt and stay ahead."

"The nation doesn't need a Marine Corps. It wants one," stated Colonel Thompson. Speaking from his experience as a Joint Task Force commander, he noted, however, "We see ourselves as small and agile. . . . We think we're small. But the USMC [U.S. Marine Corps] is as large as the entire British military." Colonel Thompson is correct. The U.S. Marine Corps is, in fact, larger than the militaries of many other countries. According to a study by Bouhan and Swartz,[19] total U.S. military spending is almost as much as the combined military spending of all the other countries in the world. (The U.S. military budget in 2009 was 42 percent of the world's total combined military budget.) So even 5 to 8 percent of the U.S. budget is significantly more than most countries can afford.

"The Marine Corps is a light infantry force, right?" queried Colonel Monroe in a lecture to an audience of Marine students. "No we're not! We're a light infantry force with a pile of logistics behind us." Colonel Monroe's observations challenge the notion that the Marine Corps is truly light and mobile. Providing a pragmatic recognition of this reality, in a 2011 memorandum by General Amos, the Commandant of the Marine Corps, he described the Marine Corps as a "middleweight force."[20]

Given that the Marine Corps is not unique in its combat capabilities and that its budget and size are by no means trivial enough to be ignored, what, then, accounts for its continued existence? There is one thing that the Marine Corps can do that perhaps no other military service in the world can accomplish: provide a rapid response team that combines all of the capabilities of ground, air, and sea combat through something called the Marine Air-Ground Task Force (MAGTF).

Flexibility in Structure and Organization:
MAGTFs, MEFs, MEUs, and MEBs

My fieldnotes from my first day at my new job at the Marine Corps are filled with an overwhelming list of bizarre acronyms of Marine Corps organizational structures. On the pages of my green four-by-six-inch fieldbook I had hurriedly scribbled notes and definitions that made no more sense than the acronyms themselves. "FMF—fleet Marine force. MEF—Marine Expeditionary Force with a division, wing and logistics group (note to myself to figure out what a division was and the definition of wings)." MEUs apparently were Marine Expeditionary Units built around a battalion and a squadron. And then there were "MEBs—Marine Expeditionary Brigades based on a regiment with air and artillery." A cryptic note next to the term MEB stated simply, "Theoretical"—it seemed MEBs didn't actually exist except on paper.

Given that I was still struggling with the most basic Marine vocabulary— such as locating the "head" to wash my hands—it is not surprising that this introduction to these complex and confusing Marine organizational structures flew right over my head. Over the coming years, I did learn the correct translations for the various acronyms as well as what a division or a regiment or a battalion was (see the Appendix for a description of these military units). And I learned that Marines had something called the "wing," or their own aviation equivalent to a division, which was structured similarly to the Air Force with squadrons instead of battalions (see Appendix). But I still struggled with why the Marine Corps would need so many seemingly complicated organizational structures.

It was not until several years later that I realized that my assumption that the Marine Corps had several different, but fixed, organizational structures was incorrect. As my research continued, it became apparent that while the Marine Corps had several official structures that existed on paper (MEFs, MEBs, and MEUs), in practice, these structures were extremely flexible, morphing from one elemental military organizational shape into new forms according to the specific purpose and mission to be accomplished.

This capacity of the Marine Corps to shape shift—to move from one organizational configuration to another or even to invent organizational configurations according to the needs of the mission—is closely connected to the Marines' view of themselves as expeditionary and mission focused, embodying ideals of agility, flexibility, and adaptability to achieve the goal. Just as individual Marines adjust their activities and functions to fit the latest new challenge, so, too, do the Marine Corps organizational structures.

A popular spoof of a poster, titled "U.S. Marine Logic,"[21] beautifully illustrates the Marine Corps' approach to organizational structure and accomplishing the mission. The image, which has been floating around the Internet, depicts a mouse in a maze. Instead of following the maze to the cheese at the end, however, the mouse has chewed its way directly through the walls to reach the cheese. The image epitomizes the Corps' rather brash, unconventional, and pragmatic approach to coming up with the most direct solution to solving any problem—making the structure fit the solution rather than vice versa.

In an interview discussing Marine Corps organizational structure, Colonel Thompson explained, "It all stems from an absolute obsession on the mission. We refuse to acknowledge it can't be done." He cautioned me not to get hung up on a specific organizational configuration. "The structure is just to justify resources, to manage equipment," he stated. What was important, he explained, was a concept called "task organization"—the idea that once Marines knew their mission or the task they were expected to perform, they would work backward, organizing the structure that made most sense to meet that task. As a result, he explained, "We don't get stuck on structure. The magic word is task organization. When we sent that BLT [battalion landing team] in to Afghanistan [referring to his earlier deployment], we sent all sorts of components a battalion doesn't usually have. We took a Force Recon platoon [force reconnaissance], fires [kinetic combat assets],[22] Intel [intelligence], Psy Ops [psychological operations]. General Mattis said 'Here is the fight' [so we gave him what he needed]."

Colonel Thompson sums up this notion that it is the Marine flexible ethos that leads to their adaptable way of operating, rather than vice versa: "In Britain, they say 'we need to operate like the Marine Corps.' But they can't. This [flexible way of operating] has nothing to do with structure. It has everything to do with our ethos."

Marine Corps organizational flexibility and adaptability are not simply random according to the latest new situation, however. Instead, the three primary organizational structures—MEFs, MEBs, and MEUs, as well as their derivatives—are variations of a core set of components that underlie all major Marine Corps organizational and functional structures: the MAGTF. The MAGTF can be seen as a set of building blocks (somewhat like a Lego kit) with four primary elements that can be rearranged and broken into parts depending on the actual task at hand: a ground element composed of Marines who are trained to fight and use combat equipment on the ground such as the infantry or tanks; an aviation element composed of Marines who maintain or fly planes and helicopters for transport and air combat; a support and logistics element composed of Marines who support the combat groups—such as logistics, supply, maintenance, communications, and so forth; and finally a command element consisting of those Marines who lead the specific organization. A fifth element—the "water" element (ships and naval crew)—is provided separately, when needed, by the Navy but is not included in the MAGTF.[23] Various combinations of these elements make it possible for the Marine Corps to conduct operations on land, air, and sea—simultaneously or separately—as the situation demands.

As an exhibit on the MAGTF at the National Museum of the Marine Corps proudly displays, the MAGTF is considered one of the Marine Corps' greatest innovations. The basic concept is that all the elements (ground, air, logistics, and command) share a combined plan and goal, working together to achieve a mission—whether leading a combined air and ground assault during the Vietnam War or delivering emergency food and water by helicopter and ground troops to victims of an earthquake in Haiti. It is this concept of working together as one self-sufficient, organic team that makes the MAGTF and the Marine Corps unique among the U.S. military services. And while the MAGTF as a general concept never changes, the particular configuration and the exact combination of pieces from all four elements varies, resulting in an organization that is constantly changing and shifting its shape, adapting to meet the particular crisis at hand.

While the task a particular Marine organization must accomplish dictates the immediate structure, a second factor that also influences the particular shape of that organization is whether or not it is in the "front or the rear." Marines and Marine units that are deployed around the world carrying out the various missions of the Corps are called "forward"—being in the front of the action. Those that are stuck back in the United States taking care of the forward operating forces are described as being in the rear, behind the action. Reflecting their amphibious identity, Marines will talk about their forward deployed brothers and sisters as being in the "fleet" (out sailing at sea), even when the "fleet" happens to be on land fighting a war in Afghanistan!

When in the rear, usually Marines return to their "home station," which is one of three MEFs. I MEF is located on the West Coast (with primary bases in Camp Pendleton and 29 Palms) and is focused, in theory, on operations in the Pacific Ocean. II MEF is located on the East Coast (its primary base is Camp LeJeune) and typically focuses its operations on the Atlantic and Mediterranean regions. III MEF focuses on operations in the western Pacific, with primary bases in Okinawa, Japan, and Hawaii.

However, even the notion of a "home station" is not so clear. MEFs, like most other Marine Corps structures, are organizational configurations that can be sent into combat too. "Desert Storm and OIF I [Operation Iraqi Freedom I]—both consisted of a MEF plus," noted Colonel Thompson. "All of these structures are just starting points. III MEF is structured very differently than I MEF and II MEF."

In principle, MEUs derive from their home station MEFs and conduct their "floats" in their home MEF's region. Reflecting this relationship, the first digit for each MEU is the same number as the MEF from which it originates (for example, the 21st MEU is tied to II MEF). However, if the Marine Corps requires assistance from the MEU outside of its home region, then that is where MEU will deploy. In the past few years we have seen "global sourcing to feed Afghanistan and Iraq," Colonel Thompson explained. "There are seven MEUs. One is permanently based in Okinawa [Japan]: the 31st MEU. It covers the West Pac [Western Pacific] and by and large has stayed there. But the other six MEUs—three on the East Coast, three on the West Coast—they used to focus on one region: the Mediterranean and Africa for the East Coast MEUs. The Pacific for the West Coast MEUs. But now they [East and West Coast MEUs] have all converged on the Middle East."

In theory, a MEB is a structure located in size somewhere between a MEF and a MEU. "MEBs, we always kind of struggled with. We don't have a standing structure," Colonel Thompson stated. "We haven't had a standing MEB since the '80s. On paper we said we'd love to have standing MEBs, but the resources weren't there." However, that may be changing. In an analysis of the Marine Corps force structure published in 2011, one of the recommendations was "resourcing five regionally focused MEB command elements."[24]

As I have illustrated in this chapter, two core components of Marine Corps identity are the ability of the Corps to adjust and be flexible and its hard, lean, austere Spartan self-image. These ideals are both tied to the Marine Corps' expeditionary and amphibious role and then reproduced through the Corps' flexible organizational structures. Structure thus is the reflection of Marine ideals of "semper gumby" and the willingness to do whatever it takes to adapt to the situation to meet the mission.

4 "Honor, Courage, and Commitment"

Instilling the Warrior Ethos at the Recruit Depots

1ST BATTALION'S PT (PHYSICAL TRAINING) FIELD IS moist and slick this September morning after a heavy rainfall on Parris Island, South Carolina, the night before. It is 0730,[1] and the shaved-headed recruits have already completed their morning run, drenched by sweat and a light drizzle that is still drifting over the field. I, too, have been up since 0430 in order to observe the recruits at the Marine Corps Recruit Depot (MCRD)—more colloquially known as "boot camp"—perform their morning physical training. At the moment, Platoon 2076 is alternating with Platoon 2075 through several rotating series of pull-ups and chin-ups on the bars to the right side of the field, sit-ups on the incline boards, weightlifting in the sandbox in the middle, and jumping jacks and repetitive step-ups on the wooden steps to the right of the field. The movements of the recruits are fairly efficient and organized, and I can tell by the bloused trousers in their boots that these Marine hopefuls have already completed the first phase of training at the depot.[2]

"Recruits to the center field! Get in line. Move it!" Standing tall and terrifying, the drill instructor (DI) barks at the recruits as he paces back and forth along the edge of the field, his green belt lying flat on a lean, taut, muscular body that dares any recruit to defy him. "Pick it up. NOW! Jones to the right. To the right! What are you—SLEEPING? This ain't no country club. No one's gonna come and give you no INVITATION. Move!" Recruit Jones, a thin, wiry eighteen-year-old from the Bronx, hurriedly moves into place to stand face-to-face with his martial arts partner—a short, stocky Iowa farm boy.

"Arms up, elbows bent—like this! You understand? Keep the fists tight. Stare your opponent in the eyes. Show you're in control. Understand?" the menacing martial arts instructor shouts out from on top of the "table"— a wooden podium in the center of the field—as she steps into the combat stance for the surrounding recruits to see. I am surprised to see a female Marine teaching the male recruits. So I lean over to Staff Sergeant Todd, one of the drill instructors for the famed Marine Corps Martial Arts Program (MCMAP) who is observing the exercise next to me. "Sure a female can teach martial arts," he responds. "If she passes the certs [certification tests], then she's qualified. Male or female doesn't matter."

"Right now we're learning ear choke," Staff Sergeant Todd explains to me as the recruits step forward and wrap their elbows under the back of the head and around the neck of their opponent. "A lot of these kids have never been in a fight, never been hit before. One big thing is that once they've been hit, they're not afraid. Confidence. Everything on Parris Island is to build confidence."

Several of the recruits slip on the muddy field. The lean, wiry instructor on the table shouts over to one of the recruits on the ground. "Bend those knees. Bend the knees! Like this. Understand?" She takes the warrior stance again.

Most of the recruits are now splattered with mud, and I shiver slightly in the rainy morning chill. Rolling around in the mud and rain is far from what I consider to be an enjoyable start to the day, so I'm a little surprised to hear, "The MCMAP program is all about the warrior ethos. The kids love it. On day two [of formal training, which begins after a week of in-processing and 'forming'], they go out there and they finally get to do martial arts. They've been here about a week. And they finally get something 'Marine like.' That's what they signed up for." Staff Sergeant Todd leans over, a look of pride in his eyes. "MCMAP really sells the Marine Corps to recruits."

I am rapidly taking photographs on my small digital camera—trying to record in pictures that indefinable but perhaps quintessential heart of being a Marine—the fighting spirit. "The program is really about selling the program. If you can make someone believe that the moves will work, then they'll be able to do it. They'll be scary," Staff Sergeant Todd continues. MCMAP isn't about doing the moves perfectly, he explains. "It's about the warrior spirit. You can tell by the demeanor. He'll go right in and he'll do it and you'll know. . . . I'll see a recruit who doesn't have the moves, but he's got that right attitude. And we'll say, 'He's got it.'"

The staff sergeant pauses, checking the field to see how the recruits are doing. Satisfied, he turns to me and adds, "Another thing we stress in MCMAP—you can be that warrior on the battlefield, but when you come off the battlefield, you become a gentleman." He points to a recruit who is leaning over to help two others who seem to be in a disagreement about the right way to do the move. "We stress that in the instructor training. We don't want that bully. . . . Strong doesn't mean being a bully; it's about being a well-rounded person. If you don't display the three characteristics—mental, physical, and moral, then you can lose your ability [to do your job right]."

What is this "warrior ethos" that is at the heart of Marine Corps identity? This attitude that changes a sloppy, shuffling teenager—who steps sleepy-eyed onto the famed "yellow footprints" upon arrival at the recruit depot—into a tall, upright, proud Marine at graduation twelve weeks later? Equally important, is it possible for a Marine to be both a "fearless warrior" and an honorable gentleman (or woman) as Staff Sergeant Todd suggests? Is there no identity crisis or internal tension for Marines—who are trained to be powerful and terrifying—to also be culturally astute or respectful of the needs of others? How has the Marine Corps managed to reconcile these two seemingly incompatible roles and identities?

Marine identity is firmly rooted in pride in the Marine Corps' history and the belief that Marines are *the* elite fighting force in the world. Membership in this elite group is considered a privilege. As I argue in this chapter, the modern-day Marine embodies a particular archetype of warrior—akin to the chivalrous white knight in shining armor who simultaneously "slays dragons" (as one of the majors in my classes poetically explained to me) and rescues damsels in distress. This ideal chivalrous Marine honors the past by adding glory to Marine Corps history—combining both strength and bravery with integrity and socially responsible actions.

In order to socialize these two identities—fearless warrior and honorable gentleman—recruit training employs two parallel approaches to imprinting core values on the souls of Marines. First, recruits are taught the importance of Marine pride and their role in upholding the legacy of Marine Corps history. Second, a values based training (VBT) program founded on the principles of "honor, courage, and commitment" is woven throughout the initial recruit training curriculum. Through these two approaches the Marine Corps has sought to reshape or at least reemphasize the more honorable aspects of the stereotypically fierce character of the classic ground combat Marine warrior that is "sold" to future recruits.

Recruiting the Courageous, Hard-Charging Devil Dog

> **Courage:** The mental, moral, and physical strength to do what is right in the face of fear.[3]

In contrast to most non-organizational cultures, where socialization begins at birth and is not optional, Marines choose to enter their culture in young adulthood (generally between ages seventeen and twenty-six).[4] Since the United States has an all-volunteer military, one of the most critical tasks of the Marine Corps is recruiting new Marines to take the place of those who have chosen not to renew their contracts, retired, died, or been injured in combat—a difficult job for a country that has been at war continuously since 2001. Consequently, the Marine Corps must invest considerable time and energy into defining who and what a Marine is: marketing this identity to potential recruits. The Marine Corps' recruiting process and the attendant marketing materials, therefore, provide a fascinating and unique window into Marines' self-image (or at least the image they choose to portray to the public).

Marine Corps Recruiting Ads

The billet of recruiting duty is considered one of the most challenging positions a Marine will hold. As Colonel Simons, who formerly worked at Parris Island, explained to me, "We put recruiters in a tough spot. Recruiters have got to make their quotas. It's often described as thirty-six one-month combat tours." Amazingly, despite the almost certain fact that anyone entering the Marine Corps in recent years will be sent overseas into one of the current U.S. wars, he added, "We tend to attract the kids we need." According to retired Lieutenant Colonel Custer, now a U.S. government civilian working for the Corps, this success is because "the Marine Corps, next to Stalin, has the best propaganda machine in the world." [5]

In fact, the Marine Corps is very proud of its ability to attract some of the best and the brightest of all the recruit pools entering any U.S. military service. According to a recent command brief (formal presentation usually in the form of a PowerPoint)[6] from the Marine Corps Recruiting Command, 99 percent of current Marine Corps recruits are high school graduates; 70 percent scored in the top three categories of the Armed Services Vocational Aptitude Battery (ASVAB) test; and all are required to pass both moral requirements (no serious drug use or police record) and a physical fitness test (three-mile run, twenty pull-ups, and 100 crunches—for males—and a maximum weight/ obesity measure). This is quite impressive considering that, according to the brief, "75 percent of the applicants in the primary target market [ages seventeen to twenty-one] do not meet USMC standards."[7]

Adapted to a contemporary media-saturated world, catchy Marine Corps television and radio ads, poster-board slogans, websites, videos, and YouTube sell a particular identity of the Marine to would-be recruits and their parents. These advertising campaigns are carefully designed to attract the current generation and use all the latest advertising analyses and strategies to appeal to the dreams, experiences, and hopes of the newest recruiting pool. A Marine Corps booklet on the recruit training process titled *Beginning the Transformation*[8] describes the new recruit generation of Millennials as technologically connected but also (in a reversal from their self-absorbed predecessors, Generation X and the baby boomers) focused on service to their country and community.

Current Marine Corps recruiting ads focus on these characteristics, merging the identities of the strong warrior with the chivalrous knight. "A couple of years ago, ads were very much game based—[images of players with] swords, out fighting dragons, going across a bridge, saying, 'It's going to be a challenge. It's going to be tough,'" Colonel Simons explained to me. According to Colonel Tate, a commanding officer at one of the Marine Corps schools, "The new recruit generation is devoted to service. [We are now focusing on] the Marine Corps as a public service. We have a new ad campaign to show that the Marine Corps is an opportunity to serve, just like Americorps." In fact, the 2013 Marine Corps recruiting website includes a page titled "Your Impact," which describes the contemporary socially responsible warrior: "Standing up against tyranny and injustice are Marines: elite warriors who courageously and honorably face down the threats of our time."[9]

Although a complete study of Marine Corps advertising and media representations (both past and present) would require a book in itself, a quick glance at a few of the more popular Marine Corps websites, videos, films, posters, and sayings reveals some of the images that the Marine Corps uses today to attract its new recruits. The first and traditionally accepted image promoted is of a physically strong and courageous fighter. In the YouTube video series *U.S. Marine Corps: Making a Marine (Parts I, II and III)*,[10] which shows live footage of initial recruit training on Parris Island, the viewer is dazzled with video clips of new recruits who learn to fight with bayonets and pugil sticks, shoot rifles, attack their opponents using martial arts, rappel down sheer walls, climb over and under obstacle courses filled with barbed wire, and survive the grueling night marches, hunger, and challenges of the Crucible (a fifty-four-hour culminating endurance test). Throughout the video, close-ups of the new recruits show faces racked with pain, sweat dripping down their brows, bulging arm muscles straining. Sound tracks include heavy panting, groans and thuds, machine-gun fire, and, over it all, the screaming voices of the DIs.

As the video displays, an equally important aspect of their culture that Marines advertise—and frequently use to distinguish themselves from the other U.S. services—is that of the tough, determined warrior who endures pain, hardship, and suffering in order to always get the job done. Thus, from the very beginning of the recruiting process, the Corps advertises and promotes its Spartan warrior ethos. Marines refer to themselves as "hard chargers" pushing through any difficulty. And they are proud to speak of themselves as being at the forefront of battle: at the "tip of the spear" and "first to fight." The following sayings, popular among Marines (and sometimes worn on T-shirts or displayed on mugs in the office and on recruiting posters), reflect Marines' attitude that pain, suffering, and endurance are virtues:

> "Pain is just weakness leaving the body."
> "Losers quit when they feel pain; Marines quit when the mission is
>   accomplished."
> "Pain is temporary. Marine pride is forever."
> "Nobody ever drowned in sweat."
> "When it absolutely, positively has to be destroyed overnight."[11]

The Marines' official "mascot" is a bulldog—mean, ferocious, tough, and scary. Marines adopted this icon after the Germans nicknamed them "Teufel-hunden,"[12] or "Devil Dogs," after the ferocious mountain dogs of Bavaria. It is an icon and nickname that Marines carry proudly; in fact, it appeared in a famous recruiting poster during World War I.

Famous classic recruiting posters have also attracted future Marines by advertising this tough, fighting Spartan image (see Figure 4.1). These earlier posters displayed slogans such as "Want action? Join the Marine Corps"; "First

FIGURE 4.1 Historic Marine Corps recruiting posters. Source: Art
Collection, National Museum of the Marine Corps, Triangle, Virginia

to fight. Join now and test your courage. Real fighting with real fighters"; and
"We don't promise you a rose garden." Indeed, in contrast to the other U.S.
services, the Marine Corps does not advertise goodies and bonuses to attract
new recruits (although college tuition, travel, and so forth are part of the benefits of joining the Corps). As Stukey writes in his unofficial guide to Marine
Corps warrior culture, "Instead the Marine Corps recruiting advertisements
tout the *challenge* of their Warrior Culture."[13]

Contemporary Marine Corps television advertisements continue to sell a
tough, hard, Spartan image to potential recruits. One fast-paced ad titled "No
Compromises,"[14] for example, shifts rapidly through images of Marines scaling

walls, driving tanks through the desert, crawling on the ground holding rifles, swooping down in helicopters, and standing tall in drill uniforms. The only words that flash on the screen are "We don't make compromises. We make Marines." Another video titled "A Path for Warriors" alternates images of a boxing match with those of Marines running, fighting in hand-to-hand combat, moving through the jungle, dropping out of helicopters, and running with their rifles across the desert. The video's text flashes on the screen: "There is still a path for warriors. Paved with sacrifice. Illuminated with purpose. Covered with sweat. . . . Proven in the field of battle."[15] These YouTube videos exemplify all of the stereotypes I held of the Marine Corps prior to my arrival at Marine Corps University, glorified through Hollywood films from John Wayne's classic role in *The Sands of Iwo Jima* (1949) to Stanley Kubrick's *Full Metal Jacket* (1987) or to even more controversial recent films such as *Jarhead* (2005).

However, there is a second, more recent identity being sold and trained into Marines. While the physically tough ground combat warrior still dominates the traditional advertising media used to attract recruits to the Corps, recent ads now market a second identity of the chivalrous knight. This identity focuses on the Marine as a rescuer and protector of justice and freedom for those who are oppressed. For example, a video released in 2012 titled "Towards the Sounds of Chaos" starts out with sounds of desperate people screaming and images of Marines running to a tsunami-strewn beach. The narrator begins, "There are a few who move towards the sounds of chaos ready to respond at a moment's notice. When the time comes they are the first to run towards the sounds of tyranny, injustice, and despair." The video ends with the challenge: "Which way would you run?"[16]

Similarly, on the Marine Corps recruiting website there is a page titled "Global Impact" with subsections titled "Worldwide Impact," "National Impact," "Community Impact," and "Your Impact." The text below the "Worldwide Impact" subsection reads: "In the face of global unrest, there is no rest for those of us at the tip of the spear. Marines take the fight to the enemy and give relief to those in despair. Against tyranny, disasters and threats to the world, Marines fight all over the world. These are not fights for which to apologize. These are fights to win."[17]

This moral dimension of the Marine, as someone who upholds the values and ideals of freedom, justice, and honor for the oppressed around the world, is one that is socialized from the day a young civilian expresses the desire to become a Marine, becoming a "poolee."

Socializing Poolees into Marine Corps Values

"Poolees" are those civilians who have expressed interest in joining the Marine Corps but have not yet been sent to boot camp. In part because of practical considerations (ranging from available spots at the recruit depots to helping the poolee prepare legally, emotionally, and physically to pass the required entrance tests) and in part because the Marine Corps does not want to invest in a recruit who is likely to quit during initial training, a potential recruit will not "ship" (be sent) to the recruit depot for six to twelve months after he or she has signed up. During this time, the applicant becomes part of the recruit "pool" and will attend weekly or monthly activities organized by the recruiting officer in order to be socialized into some of the basic aspects of Marine Corps culture. These activities range from going for training runs and swims to "meet the DI night" (where parents and poolees can come and talk to DIs about boot camp), to watching movies and eating popcorn and talking with their future Marine brothers and sisters.

"The change from being a civilian to being a Marine starts with the recruiter," explains Colonel Simons. It is the recruiter's job to prepare his or her poolees not only physically for the Marine Corps but also mentally and socially for the new culture they are entering. During this early socialization process, poolees learn some basic facts about Marine Corps history, pick up a bit of the Marine Corps language, and, most important, become exposed to a second critical part of Marine identity: the Marine as honorable—upholding the history and legacy of his or her predecessors, committed to his or her fellow Marines, and worthy of membership in the Corps.

This concept of the "honor-bound" Marine is exemplified in a YouTube video titled simply "New USMC Recruiting Video."[18] This clip provides a much more personal and lengthy advertisement than the typical thirty-second television piece and emphasizes the moral dimension of being a Marine. In this video the classic images of hard, tough Marines running through the desert with rifles, fireballs burning around them, are punctuated with another less typical image—Marines handing out aid to shy Iraqi girls and Arab villagers waving as Marines drive by on tanks. Repeating famous Marine Corps sayings, speakers narrate over the fast-paced images with familiar statements such as "Marines have always been there to take the forefront. To take the tip of the spear." However, these are interspersed with more value-laden statements: "In the Marine Corps 'honor, courage, commitment' that's what we believe." "What we do extremely well is take care of each other." "Semper Fidelis is us continuing on. . . . That's the legacy of being a Marine." The video ends with the written phrases: "Are you ready?" "Are you worthy?" and the classic Marine saying, "The few, the proud, the Marines."

As this recruiting video illustrates, while the Marine Corps recruiting process does serve the function of preparing recruits physically for the depots, an equally if not more important purpose is to socialize prospective recruits into the moral and ethical values of being a Marine. And these values are transmitted through venerating Marine Corps history and the courageous and honorable actions of the Marines who have come before.

Semper Fi: Commitment to Upholding the Historical Legacy of the Band of Brothers

Commitment: Unwavering, selfless dedication to mission accomplishment and personal and professional responsibilities.[19]

Across the U.S. military services, Marines are famous for the pride they take in the Corps and its history. This was brought home powerfully to me during a "backbrief" (a summary report back to a unit) of some of my work on Marine Corps identity and the warrior ethos. Due to the iterative and collaborative nature of my research—reflecting in great part the culture of the Corps itself—at one point in my research I created a several-slide PowerPoint for some of the Marine officers who had assisted in my research. The slides described Marine Corps identity using such categories as warrior fierceness, warrior toughness, team spirit, self-control, and leadership.

After a few minutes of looking at the slides, however, Colonel Thomas remarked that I was missing perhaps one of the most important aspects of being a Marine. "An understanding of history, of legacy, has to be added," he remarked. "Our myths, our legends, our traditions. . . . If we ever lose that sense of history, the ethos aspect of it, [we will stop being Marines]. We are really big on our legacy. We have the Birthday Ball [the annual Marine Corps ball that celebrates the birth of the Marine Corps on November 10, 1775]. We know we have icons [like Chesty Puller—one of the bravest and most decorated Marines in the Corps' history]."

For Marines, history is much more than simply a boring list of names and places from the past. As General Grant stated in a speech at one of the many Marine Corps Birthday Balls held around the world wherever there are

Marines, "To us, as United States Marines, our history is everything. [We] are following in the footsteps of some very amazing people. Marines who sacrificed, fought, and died to establish the honor and traditions of our Corps. And it's our duty to uphold the honor and to carry on the traditions of those who have gone before."

This moral obligation of Marines to live up to the legacy of their forefathers and, indeed, to add to this glorious history is illustrated in General Grant's following remarks. Addressing the Marines in the room who had returned from combat in Iraq and Afghanistan, he stated proudly: "Trust me when I tell you that future generations of Marines will recall the valor, heroism, and sacrifice that you and your fellow Marines have demonstrated over the past decade of war. You and all Americans should be proud of the legacy you've established."

In this ideology, Marines are not simply the recipients of history but the makers and shapers of history, both upholding the honor of the past and also creating the glorious history of the future. Marines remind each other daily of their bonds to each other and the legacy of the past by ending most interactions with the saying "Semper Fi"—based on the Latin *Semper Fidelis*, or "always faithful." Each Marine is in essence a part of the body of the Corps—not an irrelevant name considering that no other U.S. service refers to itself as a Corps.

This connection to a long lineage of courageous warriors creates cultural bonds of "fictive kinship" among Marines. Closely related to the veneration of one's "ancestors" is the notion that those who have gone before are not dead, but live on through the glorious deeds of contemporary Marines. Thus the honorable actions of today's Marines carry on the spirit of their forefathers. "There is this belief that we're being watched. Chesty Puller or your grandfather who was a Marine [is looking down on us]," Colonel Thomas, a commanding officer at one of the training schools, explains. "There's a spirit out there. It's not about us, it's about the institution."

This mystical, almost spiritual nature of being a Marine was brought home to me during one of my many visits to the National Museum of the Marine Corps in Quantico, Virginia. Built in collaboration with the Marine Corps History Division (the very existence of such a division emphasizing the importance of the past to Marines), the museum acts both as a chronological document of many of the Corps' most glorious moments and also as a repository of famous artifacts from the Marine Corps' past. Understanding that history was somehow central to Marine Corps identity, I made several visits to the museum—trying to figure out who Chesty Puller, Dan Daly, and Hashmark Johnson were and what happened at Belleau Woods, Chosin, and Iwo Jima. These names not only were mentioned frequently by the Marines I worked with but also served as "street signs" on the trails behind the Officer Candidates School, where I biked to work. So I figured that to understand Marines, I would have to understand their history.

However, try as I might to find the hidden meaning in the exhibits displaying Marines huddled in the trenches at Belleau Woods in World War I, dead bodies lying frozen in the icy mountains of Chosin in Korea, or the famed images of four Marines and a Corpsman lifting up the American flag on Mount Suribachi, Iwo Jima, in World War II, I just couldn't understand why these names and places mattered so deeply. Then by luck, one day as I was wandering through the museum, I ran into Colonel Moore, who also worked on the Marine Corps Base at Quantico. He was delighted to hear of my research project and noted that if I wanted to understand Marines, I had come to the right place. After some conversation about the various exhibits, he unexpectedly explained the almost religious role of history for Marines: "We sometimes refer to this museum as 'The Temple.' Inside are many items that are sacred to us from our history."

The Marine Corps' cultural ideal of upholding the honor of the sacred past, of carrying on the glorious deeds of the Marines who have gone before, is the vehicle for incorporating ethics and values into Marine Corps training at the recruit depots. And the first step (literally) to becoming a Marine is to demonstrate that one is worthy of carrying on and being faithful to the great history and traditions of the past—a twelve-week initiation ritual colloquially called "boot camp."

"Stepping in the Yellow Footprints"

Two silver minivans pull up in front of the Parris Island Recruit Depot. The time is 1930 (about half an hour after sunset), and the first of the new recruits for the evening are being delivered. Slowly a set of disheveled, disoriented teenagers, mixed with a few twenty-somethings, stumble off the bus.

"Put your feet on the yellow footprints," the DI barks as the new recruits rush anxiously to find a spot among their future colleagues. "Males up front. Females in back." The DI paces back and forth impatiently along the edge of a pavement marked with yellow footprints, set in neat rows in front of the steps to the in-processing building.

When the group of thirty or so young men and women has assembled with their feet on the footprints, the DI begins with the famed introductory speech, spoken without change or variation to each new group of "pick-ups" (as the arriving recruits are called). "As a civilian you are subject to civilian law. As a Marine recruit you are also subject to the Articles of the Corps." Pacing back and forth along the line of recruits, he begins to paraphrase some of the key Articles to which all Marines are subject:

"Article 91. Disrespect. Disrespect will not be tolerated.

Disobedience. Article 92. Do what you are told to do when you are told to do it."

The DI continues to pace up and down in front of the anxious new recruits, glaring fiercely into the eyes of any who dare to look his way. "The Marine Corps' success depends on teamwork. Starting now you will live, eat, sleep as a team. Words like 'I,' 'me,' and 'mine' will no longer be a part of your vocabulary." This is no empty threat, for during the next twelve weeks of training on the depot, recruits will have to refer to themselves in the third person, raising their hands to make even a simple request such as "This recruit requests permission to use the head."

The DI pauses, stops, and stares up and down the rows of young recruits, scrutinizing the sloppy shirt hanging over a paunchy stomach of a tense young man who can barely stop from shaking. The DI weighs this new crew of potential Marine brothers and sisters, as if to say, "Are you worthy of standing beside me under the fire of battle?"

He continues his speech: "Tens of thousands of Marines have carried on a proud tradition. As of now you will continue this tradition." He turns his head to look at a set of silver doors at the top of a short flight of stairs. "These silver hatches will represent your transformation from a civilian to a Marine. Therefore, you will pass through only once." Over the building's silver doors the following words are engraved: "Through these portals pass prospects for America's finest fighting force: U.S. Marines."

Slowly, in sets of twos or threes, the new recruits walk up the stairs and through the silver doors to begin the transformation.

Becoming Worthy of the Title
"Marine": The Recruit Depots

Honor: To live your life with integrity, responsibility, honesty, and respect.[20]

In the Marine Corps, in contrast to the other U.S. military services, the title of "Marine" must be earned. The Marine Corps expects that not all who set out to become Marines will be worthy of earning the title—which is why the silver portals over the entrance to the in-processing building refer to the new arrivals as: "*prospects* for America's finest fighting force." As General Officer Grant stated in his speech at the Birthday Ball to a mixed audience of Marines, other U.S. servicemen, and civilians, "Not everyone gets the privilege of being a United States Marine. You don't get to be a Marine just because you showed up!"

In Marine Corps ideology, during boot camp Marines are "built" or "made" from the essential "raw material" of individual American civilians. In my interviews, Marines used factory production words to describe the program, such as "The new Marine is the end product"; "The drill instructor is the maker of Marines"; "Recruit training hasn't changed over the past twenty years. We build a fairly defined product"; or "We build a basic watch."

The production process is viewed as truly "transforming" the incoming recruit into someone unique and different: a Marine. An official PowerPoint slide on the process shows a stepwise progression of "building blocks"—in each block, a new set of skills (mental, moral, or physical) is added to the training progression.[21] In conversations, these slides were referred to using a biological/genetic metaphor. "The DNA slides—little building blocks of everything that's being added and being reinforced," was the way that Lieutenant Colonel Custer described them.

In addition to the metaphors of factory production and biological generation, this transformation from civilian to Marine is also viewed as being almost mystical or spiritual—a rebirth to a new life and new future, as First Sergeant Lewis describes in the opening passages to the Marine Corps publication *Beginning the Transformation*: "It's all about character, we must train them to think, talk and act core values, then it becomes habit and your habits become your character, and your character becomes your destiny."[22]

Perhaps this is why, for many enlisted Marines, the "yellow footprints" are indelibly printed on their memories. These footprints symbolize their last physical and metaphorical step as a civilian before overcoming the challenge of becoming a Marine. For the next twelve weeks of training, recruits must demonstrate that they have the commitment, strength, endurance, courage, and mental and moral character to be admitted into the exclusive culture of the Marine Corps. Only by succeeding in this three-month initiation rite, and passing the final "test of manhood" (or "womanhood")—the famed Crucible—does the recruit earn the right to be referred to as a Marine, and to wear a symbol of the eagle, globe, and anchor on his or her collar, signifying membership in this elite fighting force.

In fact, for many recruits the rank of private will exist only on paper. By the time they complete recruit training, almost half of the newly graduated Marines will already have been promoted to the rank of private first class. While technically and legally a new recruit is in the employment of the Marine Corps from the day he or she steps into the yellow footprints, in reality the Marine Corps does not recognize the newly arrived civilians as Marines, referring to all civilians arriving at the depot by the liminal term "recruit." Their indeterminate status as neither civilian nor Marine is underscored by the requirement that all recruits must refer to themselves and other recruits in the third person.

Numerous other physical symbols also indicate that recruits are not yet Marines. For example, one afternoon as I was walking around the Parris Island Recruit Depot observing a group of disorganized recruits wandering out of the chow hall (group dining hall), one of the series commanders, Captain Frost, pointed out to me, "[Those recruits] don't wear boots yet. They still wear sneakers and have the glow strap[23] across their chest [real Marines wear boots and the glow strap around their waist]. When they do go to chow hall for the first day, everyone knows they are in forming [the initial stage of training]."

Major Wagner, an officer who had spent a tour of duty in charge of recruit training programs, explained to me, "[It's a] visual thing—tells you where the kids are in training. At the beginning they don't blouse their trousers. That's a privilege that they are given at the end of the first phase. They wear their trousers rolled up. . . . Hair cuts in the third phase—they get to have a longer cut—get to wear a 'high and tight' [a Marine Corps cut where the sides of the head are shaved almost bald and the top has a short flat layer—leading to the nickname 'Jarhead']. These are little signs that give a sense of acceptance into the club."

Marines have a number of sayings that emphasize this identity as an elite, exclusive warrior group that does not grant membership in "the club" to just anyone. An exhibit of famous posters at the National Museum of the

Marine Corps displays some of the more famous ones: posters titled with such catchy phrases as "If everybody could get into the Marines, it wouldn't be the Marines"; "A few good men"; and the famed slogan, "The few, the proud, the Marines." To emphasize the achievement of those recruits who do become Marines, during my field visit to Parris Island, a bus parked in front of the parade deck[24] during the motivation run (a traditional final run the day before graduation) displayed the banner, "Earned not given."

However, as I soon discovered during my field observations at the Parris Island Recruit Depot, to be worthy of the title "Marine," recruits had to do far more than simply demonstrate athletic strength or the ability to shoot a rifle.

The Moral Transformation

During a week of field research at the Parris Island Recruit Depot in the fall of 2011, I had been placed on a hectic daily schedule in order to observe all phases of the training. I watched the new arrivals step into the famed yellow footprints; observed Marines shooting rifles, crawling through a barbed-wire obstacle course, and practicing choking techniques in the MCMAP training program; and even received a personalized class on water survival training (which included a demonstration of an instructor tied at the feet and hands jumping from a twenty-foot diving board and wobbling his way to the edge of the pool). All of these activities had filled my stereotypical expectations of a tough, physically demanding training program designed to create Marines who were prepared to "locate, close with, and destroy" the enemy.

However, as I quickly discovered, recruit training was far more than an enormous sports program. A meaningful amount of the recruit's time was spent filling out forms, practicing drill, going for medical checkups and haircuts, and, most surprising to me, sitting in classes such as Marine Corps Customs and Courtesies, Core Beliefs, Ethical Decision Making, Substance Abuse, UCMJ (Uniform Code of Military Justice), Adultery, Personal Conduct, Operational Culture, Marine Corps History, and Combating Trafficking in Persons. My review of the depot's twelve-week recruit training schedule[25] indicated an average of one to two hours per day devoted to classroom instruction, as well as another hour per day to SDI time (mentoring time with the "Senior Drill Instructor"). As the titles of the courses listed above suggest, a large percentage of these classes focused on teaching central Marine Corps values to the recruits—values that are emphasized and reinforced during SDI time. Underscoring the importance of Marine Corps history in developing

Corps values, recruits received six separate classes on the history of the Corps. (The only other class given as much time on the calendar was Combat Care, or first aid, a six-part course).

Given the extremely tight recruit training schedule (broken down into fifteen-minute intervals during each sixteen-hour day), why would the Marine Corps dedicate approximately 80–100 hours to classes on ethics, Marine Corps traditions, and core beliefs? According to Major Wagner, an officer at Parris Island, "We're not here to teach how to knock a door down and clear a room. We're here to teach basic Marine Corps values." Echoing a similar sentiment, in a discussion about recruit training, Colonel Thomas stated, "Character development and discipline is the most important part. Many people think it's the physical but it's not. Esprit de corps—it's about teaching them the history of the Corps, pride in the Corps." He continued, "People see the physical events that the recruit has to do. But that's not what's most important. What's most important is the mental, moral transformation."

In an interview with three recruits, they, too, expressed the notion that recruit training was not about developing physical skills, but rather was a personal moral transformation.

RECRUIT A: After being here this recruit has been humbled.

RECRUIT B: You become a not entirely brand new person, but you become a better person. You think about others more and are less selfish.

RECRUIT C: This recruit thought boot camp would be a lot more physical. This recruit found that it is 99 percent mental.

RECRUIT B: This recruit believes that it's all mental. When recruits think, "I can't do this" then the day is very hard.

Values Based Training (VBT)

This mental and moral transformation is developed through a values based training program. Because VBT is seen as foundational to creating an ethical Marine who displays the values of the Corps, it is woven throughout the recruit training program. VBT is taught using a combination of formal lectures in the classroom, discipline, role modeling, and discussions with the drill instructor. All of these approaches rely on one key figure in the training: the DI. As Colonel Simons explained in an interview:

The Drill Instructor School is a very tough school—three months, very rigorous. We are teaching them very nuanced types of ways of instruction. It's tough to teach values based training. How do you teach honor, courage, and commitment? What we've found is what we're teaching was OK, but it's how the instructor delivers it.

The DI is viewed as providing the moral and personal role model of the ideal Marine. "The model of the DI, their behavior, that should be the epitome [recruits] are trying to achieve," noted Colonel Ingram, a commander at Parris Island. According to Major Wagner, "Modeling—you should be conducting yourself the way you want your recruits to be. That's why the DIs don't smoke in front of the recruits." "It takes character to be an instructor. You need character to teach it," stated Sergeant Major Clancy.

Living up to this role as the perfect Marine is no easy task, as Staff Sergeant Hanson, one of the DIs at Parris Island, described: "[There's this] whole passion bit. This job is the most brutal job I have had in my life. . . . You have to have the endurance. You're the one who is running. You're that mythical, legendary DI that is everywhere. You have to have the mental, the physical endurance. . . . It's really a lifestyle we live. That's the thing. You can't fake it down here. There are some DIs who aren't doing things that are ethical and the recruits see that. There are very few that do things that are unethical, but you can ruin it for all of us by one action."

Serving as a role model, the senior DI plays a specific role in VBT through his or her dedicated hours of SDI time with the recruits. Colonel Thomas describes this role:

Values based training is taught by the senior drill instructor. The Senior DI is the father figure, someone the recruit can go to for advice. The Senior DI will sit down, take off the hat, and give them a scenario—say "You find $20 on the ground. What do you do with it?" Their job is to guide that discussion. The recruits will come from all different backgrounds [and offer different answers]. The DI will say, "Here is what a Marine will do."

In addition to role modeling, the second approach to instilling values in recruits is through teaching discipline and self-control. "We teach core values through repetition and example. We're building a Marine who is disciplined," Colonel Thomas explained. Lieutenant Colonel (ret) Custer elaborated, "All we're trying to do is force everyone to do it the same way so that we have a standard. We can't crack open the Marine's heart and see if you have a brand but we can say 'You've been through 12 weeks of training and you've been given this.'"

One of the key methods to teaching discipline is drill. "The purpose of drill is not so you march pretty. It's to instill instant obedience to authority, discipline, and teamwork," Sergeant Major Clancy, a former DI, added. "You're probably not going to win a new war with drill but you'll have a Marine who can do what we need him to do." The importance of drill can be seen not only in the frequency that recruits practice it, but also in the name of their instructors, who are not called "recruit instructors" or "Marine instructors" but "*drill* instructors." Drill is valued not only because it instills discipline but also because it upholds Marine Corps traditions. "The Drill Master is a prestigious billet," Captain Frost pointed out as we stood watching the recruits practicing various formations on the Peatross parade deck. "It is a neat billet for the drill master because they really are the keeper of the tradition. Whatever that drill master teaches will spread out into the Corps."

At first, it seemed strange to me that ethical decision making could be taught through discipline, drill, and obedience. However, as I discussed the issue further with Colonel Thomas, it became clear that recruit training was truly not another sports training program, but a complete reenculturation program, requiring absolute acceptance of the new culture by the recruit. "It's behavior modification," he stated. "I don't care if you're from Georgia and you hate blacks. The Marine Corps does not tolerate racism, and you are going to adopt our values. We'll drop a recruit that can't adhere to the values of the Corps." He paused and then added, "That is a serious example. Our first tool is to recycle. One of the most painful things you can do to a kid is drop them back two weeks." He noted, "The Army wants to modify their training to each kid's individual character. We don't do that. They have to adapt to us." In fact, he added, "One of the reasons we have difficulty attracting [certain minority] populations is we require total assimilation."

These are no hollow statements. The Marine Corps is so serious about ensuring that graduating recruits have committed to the Marine warrior ethos and values that they will actually fail a recruit on the basis of his or her inability to adopt the Corps' ideals. In discussing the requirements for graduation from boot camp with several Marine officers, I listed all the formal tasks that a recruit must pass: the physical fitness tests, the water survival test, rifle qualifications, MCMAP (martial arts program), academics (the formal classes described earlier), the battalion commander's inspection, and the culminating test, the Crucible. "But," Colonel Cole, one of the senior commanders, interjected, "a recruit can do all these things and still not graduate to the outside forces if he does not adapt. These are the requirements, but if that young man or woman does not demonstrate core Marine Corps values, we have the moral responsibility not to send them to the fleet [the general Marine Corps forces]."

Staff Sergeant Hanson narrated a story that illustrated just how seriously the Marine Corps views commitment to their core values.

> It's all about the intangibles—pride, integrity, maturity. We're expecting an eighteen-year-old [to demonstrate] maturity. Integrity is a huge one. I had this kid—he had the pride, he had the commitment. But he wrote his platoon number up inside the [bathroom] stall.
>
> And we asked, "Who did this?" And no one volunteered. [Then] someone finally told.
>
> He was the best recruit I had and I asked him, "Why did he destroy government property?" and he finally "fessed up" [that he wanted to show pride in his unit!]. And so we ended up recycling him. And he was broken [by it]. But I saw him in the chow hall later and he said, "It's the best thing I had to do."

How are these intangibles evaluated? After eleven weeks at the recruit depots, the Corps requires the recruits to pass one final test to ensure that they are ready. This test, the famed Crucible, challenges the recruits mentally, physically, and psychologically.

The Crucible begins with a night hike. Over the next two and a half days, the recruits receive only two periods of four hours of sleep and four MREs (military dried food rations). Despite being tired, hungry, and physically exhausted, the recruits must succeed in completing ten major events and passing through twenty-eight stations—a test of their ability to make decisions and act courageously and ethically under duress. Many of the stations are named "Warrior Stations," bearing the names of previous medal-of-honor Marines. Significantly, the heroic namesakes of these stations are not highly ranked officers but courageous enlisted Marines—Private First Class Garcia, Lance Corporal Noonan, Private First Class Anderson.[26]

Each station is made to replicate a problem or crisis that these courageous Marines had to face: Private First Class Garcia's leap, Lance Corporal Noonan's casualty evacuation, Private First Class Anderson's fall. As the recruits arrive at the stations physically and mentally exhausted from the strain of the exercise, they are required to read about the history of the warrior whose name is on the station and then undertake a challenge simulating the physical and moral dilemmas each warrior faced. At the end of each challenge the recruits sit down with their DI and discuss the challenge, how their team solved it, and the ethical and practical lessons they learned. Through the heroic examples and lessons of their predecessors, recruits learn to look to the past to provide practical and moral guidance for the future.

Thus history is linked to ethics, courage, and pride for the recruits, completing and reinforcing Marines' identity and responsibility as members of a long brave and honorable lineage dating back to their origins in 1775. As the warrior stations also reveal, the Marine reverence for the past has a practical as well as ideological purpose. Marines look to their predecessors as role models, whose courageous actions and decisions can provide guidance to help them in solving the moral and ethical problems of today.

* * * *

For eleven weeks now Platoon 2076 has been crawling through barbed-wire obstacle courses, taking classes on emergency first aid, learning to shoot a rifle, scrubbing floors in their barracks, studying Marine Corps history, fighting each other with pugil sticks, and discussing ethical situations with their DIs. Their culminating test has been a two-and-a-half-day set of challenges punctuated with only four instant meals and two four-hour periods of sleep: the Crucible. After fifty-four hours of rappelling down walls, working as a team to carry each other over deep pools of water, climbing down steep walls, and rescuing imaginary injured Marine brothers and sisters, the recruits walk tired, hungry, and sore toward the statue of the raising of the flag on Iwo Jima on Parris Island.[27] Over a loudspeaker, the chaplain offers a prayer. These will be the last words that the recruits hear before receiving the symbol of the eagle, globe, and anchor—a sign of their acceptance into the Corps and their new status as Marines.

> Fifty-five hours ago we stepped off on our Crucible. We are tired, we are hungry. We are sore. But we are here. We are here to join with those who came before us and to claim the title of United States Marines. . . You forged us into a team to prepare for this day when we join the band of brothers and sisters known simply as the Corps. We ask that you continue to watch over us with your grace that we may live up to the legacy of the eagle, globe, and anchor.

A drill instructor steps up to a muddy-faced recruit and presses the eagle, globe, and anchor into his hand: "This represents our foundational values of honor, courage, and commitment. The Marine Corps is a way of life. Don't ever forget that. We welcome you into our ranks. You will forever be a United States Marine. Semper Fidelis."

One week later these new Marines will graduate. Their parents, girl-friends or boyfriends, family, and friends will arrive to watch them march proudly with their DIs for the very last time. Colonel Simons describes this last, important ritual for the new Marine. "For most parents, it's the first time they've seen their kids in twelve weeks. They've put on muscle, gained weight, lost weight. They're looking pretty spiffy.... For many kids, that day they graduated boot camp—except for the day they got married and their kids were born—is the greatest day of their life. The kids and their parents are there, tears streaming down their face. Talk about boot camp as a mystical thing. It really is a transformation."

5 "Tip of the Spear"

Leadership in the Corps

I think you walk a fine line when you start to embrace emotion too much.
Emotion is irrational and you are going into an environment that must
be driven by reason. "We lost one guy today because of X"—how you feel
about it clouds that. For example, nobody wants to go out and shoot up a
car full of women and children. But it is a very possible and probably even a
very likely circumstance that our battalion is going to experience. Suppose I
had all my conditions set [for a vehicle entry point—a checkpoint in front of
the base]. And some people in a car crossed the first line. The second line.
The third line. Now they're in our kill zone. My Marines kill it. I go look in
that car and there's nothing but dead women and children.

[So this is what I would say to my Marines]: "Okay hey good job. You
did your job. You protected them [the Marines and civilians on the base].
The people in the car crossed the boundaries; you didn't. I put you in this
position; you just did your job. . . . The ROEs [rules of engagement] say that
we have to do this. You followed your ROEs."

Setting them up for success so that they understand *why* these steps
were right; that's preparing them psychologically. . . . Then they have a clean
conscience—that it was a righteous shooting. They can come back and they
can sleep well at night. That is taking care of that Marine.

—*Lieutenant Adams, recently returned from deployment in Iraq and
training his platoon to return again*

AMBIGUITY. INSUFFICIENT INFORMATION. A CAR RAC-
ing through the gate at night. The need to make a split-second
decision that could cost a young Marine and his buddies on base their lives.
In my interviews with Marines returning from deployment, a constant theme
in our discussions was the challenge of understanding the situation and making critical decisions quickly. Some decisions were successful; others less so.
Sometimes someone died as a result.

Equally if not more important in our discussions was the role that the
Marine officers played in "taking care of their Marines"—helping the young
Marine cope with these decisions and the ambiguous, not always simple outcome of the situation. Throughout my interviews and research, it became
apparent that one of the very most important values in the Marine Corps was
leadership: the ability to be decisive; to have sound judgment; to serve as a
role model and example for others; and to mentor and take care of the young
Marines in one's unit. As Captain Rhodes, a company commander in 1/7 preparing for deployment in Iraq, stated, "All Marines want to be a leader." The
importance of leading is also echoed in the Marine saying quoted to me by
Lieutenant Nance, a platoon leader for 1/7: "The best naval officers are at sea;
the best Marine officers are in command."

Although all Marines must receive certain basic skills training and demonstrate mental, moral, and physical proficiency, the process for selecting and
developing a Marine officer is significantly different than that for training
newly enlisted Marines at the recruit depots. While young enlisted Marines
learn, at least for their initial years in the Corps, to be unwavering follow-
ers, officers—including Marine non-commissioned officers (NCOs)—are
selected to become leaders. Yet leadership, in contrast to shooting a rifle or
performing drill, is not a simple skill that one can train a Marine to perform.
The very characteristics of Marine Corps leadership—described by the acronym JJDIDTIEBUCKLE (judgment, justice, dependability, integrity, decisiveness, tact, initiative, endurance, bearing, unselfishness, courage, knowledge,
loyalty, enthusiasm)—require both character and intellectual skills that can
only be developed over time.

In the previous chapter, I described how enlisted Marines are initially
trained at the recruit depots. In this chapter, I explain the training and educational processes of selecting and developing a Marine officer at Officer Candidates School (OCS) and The Basic School (TBS), as well as the important
role and relationship between commissioned officers and enlisted Marines.

Many volumes have been written on leadership, both by the Marine Corps and by specialists in the field. My goal here is not to repeat or summarize this extensive research, but to illustrate how Marine Corps leadership qualities and ideals are intimately tied to core Marine cultural beliefs and roles. Marines value leadership qualities of decisiveness and adaptability, humility and the willingness to learn from mistakes and "on the ground" experience. They expect their officers to "lead from the front" and to "take care of their Marines." These qualities are, not unexpectedly, directly linked to the expeditionary role of the Marine Corps, its bias for action and mission success, its values of "every Marine a rifleman," and the emphasis on "the guy on the ground" described in chapters 2 and 3.

Although some of these qualities can be developed in new officers, in contrast to the recruit depots where, as Colonel Simons stated, "the DIs [drill instructors] and I will continue to train you even after you give up on yourself," at OCS, candidates must demonstrate clearly that they have the leadership potential to be an officer in order to join the Corps.

"The Effectiveness Culture": Selecting Officers at Officer Candidates School

Four Marine officer candidates are standing in front of a fifteen-by-twenty-foot pool of water, debating. A dark-haired young man in the middle, who appears to be the leader, is pointing to a large rope on the ground. Suddenly, two of the officer candidates move over to an enormous metal barrel in front of the pool and attempt to tie the rope around it. The third candidate begins to shimmy carefully up a wooden pole that hangs over the water. As I peer down from a twenty-foot metal observation platform above the Leadership Reaction Course (LRC) at the Marine Corps Officer Candidates School, I watch the candidates attempting to throw the rope over the wooden pole. The rope curls up wildly as the candidate on the pole stretches to catch the rope, but fails. Again the team on the bottom throws up the rope. This time the young man on the pole reaches dangerously farther out over the water, holding on with only his two feet and one hand. Again the rope falls into the water.

Major Sule, the exercise leader, who is standing next to me, watching the various officer candidate teams in the different metal scenario cages below, notes the increasing exhaustion of the officer candidates going through the barrel scenario. "For the past forty-eight hours they've done a bunch of foot movements. We've done everything we can to give them only six hours of sleep. Last night for seven hours they had to stand watches as well. They're fatigued." The rope curls up in the air one more time, and suddenly the candidate on the pole slips, half falling, half shimmying down the pole to splash in the pool. "You show basically your true colors when you're tired. So that's when you show your best or worst side," Major Sule adds.

The three candidates rush to their fallen teammate. Two of the candidates reach out to their now rather wet colleague, who is shivering in the chilly October air. The dark-haired leader appears frustrated, but smiles congenially and pats his wet colleague on the back. Then, without pausing, he motions quickly to the barrel as he heads for the pole.

Major Sule looks at his watch and notes that there are only four more minutes before the exercise ends. It seems unlikely that the team will be able to complete their seemingly impossible task of moving the barrel to the other side of the pool without it touching the water or the sides of the pool in less than ten minutes. He motions to me to observe as the leader below hurries to shimmy up the pole. "The biggest evaluation here is decision making, communication skills. It's about leadership—overall how you did. Only 20 percent actually complete the problem. So it's really about the ability to make decisions. There should be only one person in charge of making decisions. You should see that. If you don't, you've got a problem."

Down below, the rope swings up one more time. This time the leader on the pole leans over and manages to catch it. Hurrying, he tries to wind the rope around the pole. He pulls but the rope catches. Down below, his teammates attempt to lift the barrel to ease the tension on the rope. The leader gesticulates to the right, trying to communicate that they need to move over.

"The question is—are you risk averse or not?" Major Sule continues. "Are you comfortable with not having all the information and making a decision instead of sitting back and waiting?" The leader is now shouting to his colleagues as he throws the end of the rope down to them. "You don't have to complete the problem to get a perfect score," the major adds. "The question is—if he tried something and it didn't work, how does he react? It's not about the outcome but how he makes decisions without all of the information. None of the problems are intuitively obvious how you will do this. Even if you can see a solution, you don't necessarily have the materials to do it."

The end of the rope below swings wildly over the pool, too far away from the candidates below to reach. The leader pulls up the end and tosses again.

There is only one minute left before the exercise ends. Major Sule walks to the edge of the barrier and leans over, looking for the team's DI, who is waiting by the side of the metal caged-in, boxlike exercise area. He turns to me one last time and elaborates, "Do you have the potential to do it? At OCS we're looking for leadership potential. It's about making decisions and being flexible. Innovation at the individual level."

Suddenly one of the teammates below grabs the rope and begins to pull. But at that moment the drill instructor steps out. The exercise is over. Tired, frustrated, and wet, the candidates stand respectfully to hear the DI's feedback on their unsuccessful efforts to solve an impossible problem.

It took several months after my arrival at the Marine Corps to begin to grasp the truly different worlds in which the Marines and I lived. In my world, as a scholar and professor, decisions and conclusions were something one made carefully after analyzing all the data, painstakingly collecting as much information as possible, and evaluating varying possible interpretations and explanations. In my world, submitting articles rapidly for publication without conducting adequate research was likely to get my paper or article rejected and, in extreme cases, could result in public ridicule or expulsion from my job.

In the world of the Marines, however, the reverse was true. While a mistake, even a fatal one (a realistic possibility), was forgivable, the inability to make a decision and its corollary, "paralysis by analysis," were dreaded conditions to be avoided at all costs. As Major Sule explained to me as we observed the leadership reaction course, "Being indecisive in this organization—if you don't make a decision, you get run over." Colonel Irons described the Marine Corps' predilection for action to me this way: "Here's the real issue. You've got a mission and you've got six hours to develop a plan and execute. So you don't have time to call back and ask someone at home about the details of the plan." He continued, "We figure out this stuff while the bullets are flying. We are very much an organization with a bias for action. Action is so much a focus we have bias against thought. It's 'Do something lieutenant' and then we'll figure it out." In a later conversation he reiterated, "Remember who we are. We are right here!" Putting his hand right in front of his nose to emphasize how close Marines are to the action, he stated forcefully, "We execute and we are executors."

"Marines are driven by a need to be useful. They're the most 'can do' of all the services," Dr. Blake, a professor at Marine Corps University and a retired colonel, explained to me. Captain Rhodes observed in an interview about his deployment to Iraq, "The general Marine ethos is about doing and acting." "It's hard to justify things without results," noted Major Bennett, a former member of an Iraqi military transition team. Speaking to a planning team, Lieutenant Colonel Sampson similarly emphasized the importance of success in accomplishing a mission or task: "Let's get the mission accomplished. It's not what's on our collars [referring to rank]; it's not what our billets are. It's about getting the job done." In another planning exercise, Colonel Samuels echoed a similar sentiment: "I'm a simple man. I'm an infantry guy. 'Give me a job, boss, and I'll get it done.'" Perhaps Lieutenant Colonel Orr, one of the faculty advisors at the university, best summarized the Marine Corps bias for action, however: "We are kind of in the 'effectiveness' culture."

This ability to respond quickly and stay in front of new and changing situations is seen as a critical component of leadership. Lieutenant Colonel Orr continued, "Good teams stay in front of the problem. They see things as they are emerging." In my interviews, Marines not only expressed the importance of being able to stay "in front of the problem" but also saw themselves at the forefront as leaders both in battle and in a crisis. One of the more colorful Marine sayings provides a very visual image of this notion of being in the thick of the crisis, directly pointed at the problem. As Dr. Green explained, Marines speak about "being 'at the tip of the spear'—where it's happening."

This cultural notion of being at the forefront—able to respond quickly to new and emerging problems—is tied to the nature and purpose of the Marine Corps as an expeditionary military force. Not surprisingly, then, the ability to have good judgment and make decisions quickly under ambiguous and stressful situations is an important measure of whether candidates make it through OCS. In fact, for the Marine Corps, OCS is not a school to train future officers; it is a screening program to determine which candidates demonstrate the necessary qualities of leadership, showing the potential to become an officer. "OCS really is a ten-week job interview," Lieutenant Colonel Gilmore began in an interview about his role as a former commanding officer there.

> We're evaluating them, particularly when you talk about integrity, leadership. OCS is all about gauging leadership. Half of the candidate's score is based on leadership. We use peer evaluations—talk about being brutal. [We want to know] "Does that kid have what it takes to lead Marines?" We put them in all possible situations. We test them physically—can they work through the haze of being tired all the time? They're supposed to have eight hours of sleep but it doesn't happen. Decisions—we make them make decisions. They get a

two-day leave after two weeks. "What am I going to do? Take a drinking trip or study?" Mondays are pretty stern. We see how they handled time. . . . At OCS we're not trying to teach them anything. We're trying to evaluate them. We give them too much too fast. What do they do with it?

Although officer candidates at OCS and recruits at the depots do share a few similar experiences—undergoing physical challenges such as obstacle courses, learning some basic martial arts and Marine Corps history, and undergoing some kind of endurance course—that's where the similarity between the two schools ends. Like recruits at the recruit depots, individuals who hope to become Marine officers are given a liminal title (candidate) until they have earned the title "Marine." However, unlike the recruits, who are at least legally (though not socially) a member of the Marine Corps while at the depots, Marine officers are not granted a commission (a legally signed presidential appointment as an officer in the Marine Corps) until *after* they graduate from OCS. The distinction is deliberate.

While the Marine Corps is willing to invest extra time and training in an enlisted recruit to help him or her succeed, at OCS the goal is to select those candidates who show potential as leaders. "We call it '*screen in*' not '*screen out*,'" Major Sule told me. "At OCS the question is, 'Do you have the potential to do it?'"

The attrition rate at the recruit depots is about 7 percent for the males (double that for the females). In contrast, up to 25–30 percent of the males and 50 percent of the females never make it past OCS. In the case of the females, it is often because the program literally "breaks" the women: since women are expected to accomplish all the same tasks as the men, many return home with stress fractures, their smaller bodies and frames unable to carry the heavy packs or climb the same courses. Unlike the other services, the Marine Corps does not make significant adjustments in the physical requirements to become a Marine according to gender. The Corps' view is that if "every Marine is a rifleman," then every female Marine should be able to carry the load of a male Marine. And if an officer needs to command respect, he or she needs to be capable of doing everything that the Marines following him or her have been asked to do.

So what are the characteristics that make a candidate successful—the qualities that the Marine Corps views as critical for a new leader? According to Gunnery Sergeant Puller, one of the DIs at OCS, "There's not a science to it. Not a spreadsheet that spits out who should be an officer."

Major Swanson, who was working with Major Sule at the time of my trip to OCS, described the qualities of leaders that they were looking for in the candidates. "Decisiveness. We give enough information to make decisions. The weaker candidates don't have the strength to impose their view . . . they are timid and not able to make decisions and lead by example."

"We're not looking for tactical prowess," Major Morris, another member of the OCS training team, added. "We're looking for the ability to make a decision."

"It's about how they are going to react," Gunny Puller continued. "We put them in squads and assign their peers as squad leaders. When it rotates, they are now the squad leaders. Perhaps they don't do so good and they can't bounce back. It's how they react to failure, not if they pass or fail."

Leadership is not just about the ability to make decisions or demonstrate authority, however. As Gunny Puller was quick to note, another quality that is critical is "teamwork, unselfishness. The individual who helps me as the leader, who goes out of his way to help the person that's struggling. We put a large value on how they treat each other."

"An important quality of a leader is followership," Major Morris added.

"Yeah," the Gunny continued on. "The candidates figure each other out very quickly. They hurt each other for the wrong reasons. Maybe a candidate's not giving someone the benefit of the doubt and decided he's a failure. But if he's trying to do it and you're not giving, you're also being evaluated. Everybody's being evaluated. . . . We evaluate how you take criticism. Some candidates come right back, with 'you're all wrong.'"

"It's the intangibles, things that are inside—minds, hearts, and souls," Major Morris explained. Then she added, "One thing that's an absolute line in the sand is an individual's integrity. They're out if they lie, cheat, or steal."

At the end of ten weeks, after being evaluated twenty-four hours a day by the officers, the DIs, and one's peers, approximately 50–75 percent of the candidates to become a Marine officer will have completed the course successfully. These men and women will graduate and be offered a commission. Since a four-year college degree is a requirement to become a Marine officer, some of the graduates will delay their commission for a couple of years until they have completed their schooling. Some will decide that despite completing OCS, they are not interested in becoming a Marine officer. At that point, as Lieutenant Colonel Gilmore remarked, "When a kid at OCS says, 'Ah, I'm not sure I want to be here'—they're out.

We don't want an officer who's not sure this is for them, or unclear about their leadership."

And some will return home for a few weeks or months, say good-bye to their families and friends, and then begin the final journey to become an officer: at The Basic School (TBS).

"Where the Rubber Meets the Road": Creating Officers at TBS

In most Marine officers' eyes, The Basic School is where their career as a Marine begins. According to Colonel Stacy, who had once commanded a battalion at TBS, "OCS is just scratching the paint to see if it sticks. It doesn't make Marine officers. This [TBS] is where we make Marine officers." Clearly proud of his school, he continued, "TBS is the flagship institution of the Marine Corps. It's a unique place—both to the Marine Corps and the other services. TBS is the reason Marines are unique. Only Marines have TBS."

The Basic School is indeed unique. For six months, every single Marine who is a newly commissioned officer is trained eighteen hours a day, not only in the tactical skills required to be a Marine—from shooting a rifle to working with tanks to night land navigation—but also in decision making and leadership skills required to lead Marines. As a result, unlike the other U.S. military services, every single Marine officer will have received exactly the same training regardless of his or her future military occupational specialty (MOS). "There's been a certain effort to replicate TBS in the Air Force, to standardize something between commissioning and MOS, but not like TBS," Colonel Stacy explains. "The Army has looked, but they loaded on MOS more and it's only six to nine weeks."

Because of the time involved, TBS is a huge investment in each officer, as Colonel Stacy observes: "It's a six-month program—it takes six months out of the life of the Marine. The Marine Corps buys into TBS at a price, in terms of manning, the kinds of people that are here. The cost of TBS is not only time but the instructors. Everybody who was anybody has spent time here teaching." The Marine Corps obviously sees this cost as a worthwhile investment. TBS has been making officers for over fifty years, and even during the heavy deployment cycles to Iraq and Afghanistan, the Marine Corps did not cut down on the number of days or hours of training at TBS.

The value of TBS, however, is not only that it provides consistent training to every single Marine officer. From a cultural standpoint, TBS experience bonds all officers together in the same cohort. For their entire career, officers will define themselves by the year and company with which they graduated from TBS. Throughout their career, as they are shuffled around the Marine Corps, they will recognize other officers who graduated from the same cohort, creating a bond that spans time and place. TBS, in fact, is one of the reasons that despite the immense fluidity of personnel and the enormous geographic span of the Corps (literally around the globe), for most officers, the Marine Corps feels like a small and cohesive community. As retired Colonel Sorenson—a faculty member at one of the Marine Corps schools—notes, "The officers' experience for all Marines is the same—across history and time and locations. The experience bonds us together."

The small world of the Marine Corps officers was demonstrated to me in an unexpected conversation with three lieutenants who had coincidentally graduated together from TBS two years earlier. All three had just returned from deploying to Iraq, and all three were preparing for another deployment as platoon commanders within the same battalion. I had been interviewing them as a group about their experiences in Iraq. Unaware of their close familiarity with each other since they had attended TBS together, I was surprised when the conversation turned to their days at TBS and the experiences that bonded them.

LIEUTENANT ADAMS: We were at TBS. We chose, probably 95 percent chose to be there, and you wanted to learn. You wanted to learn from each other, you wanted to learn from the instructors, you just wanted to sponge everything. But it was . . . there were some suffering moments. And there were some moments that you pushed yourself. . . . I pushed myself harder the . . . what was it the fourth week? Was it the eight-day phase? [turns to his fellow Marines] You know . . . the patrol phase. You know I had eight hours of sleep in those eight days, and four of them, four of them came on the last day because we had a live fire [using real weapons with live ammunition]. And you know it was freezing conditions. Every time we went out it was raining.

LIEUTENANT BEYER: It was raining, 30°.

LIEUTENANT NANCE: Or snowing or it was miserable, a lot of misery.

LIEUTENANT ADAMS: But it was some of the best times of my life, though.

LIEUTENANT NANCE: That was probably one of the most crucial . . . just a lot of sucky times, just suffering times. And everybody does it together and you just enjoy it 'cause it sucks so bad. I don't understand it but that's kind of, it's huge.

"TBS is really about the culturalization process of making an officer," a coworker of mine, Major Neal, recalled of his time there. "How you get indoctrinated into what the Marine Corps is about. OCS is part of it—you get a lot of lore and history. But by the time you get to TBS that's when they start impregnating you with what being a Marine is all about."

Practical Experience

Although cultural cohesion is an important result of training all Marine officers together, one of the most critical functions of TBS is to give the next generation of officers genuine "hands on" experience in both leading Marines and operating combat-related equipment. For Marines, there is no substitute for "having been there and done that." This emphasis on doing and experiencing is reflected in various Marine expressions. Marines talk about the "ground truth." According to Dr. Green, this is the idea that "you have to go there and see it for yourself on the ground." He continued by noting, "It's all nice to talk about a thing. But the truth is all in where it happens—'Where the rubber meets the road.'"

The critical importance of receiving practical hands-on experience for leadership was explained to me by Lieutenant Beyer, one of the platoon leaders mentioned previously. In discussing why all Marine officers had to go through the same basic rifle training as enlisted Marines, he observed:

I think the good thing is you really understand what a Marine rifleman's capable of. Now fifteen years from now, who knows? We might have like ray guns or something. But at least you have a very solid grounding on like "where the rubber meets the road" . . . that grassroots knowledge. So when you say, "Hey Marine X, stand post (guard) here," you can say that with a pretty high level of confidence because you know that if you were in that position [you could do it]. You have a solid founding. . . . And you're not going to have to depend on other people explaining to you what a Marine is capable of. That is a tough position, to be a leader and have someone explaining to you. Or having to observe for a significant time of your command what your guys you're in charge of are actually capable of doing.

The discussion continued:

> LIEUTENANT NANCE: You have credibility in your occupation. I think that's a powerful thing. You know that you have no problem telling somebody in your platoon, "Hey, you will not shoot the weapon like that!" Because you know as well as anybody that that's not safe or that's not a relevant deployment method. As a new lieutenant stepping into a platoon if you aren't a hundred percent confident in your own proficiency it's hard to supervise somebody that's more proficient than you.

> LIEUTENANT BEYER: Confidence, live fire maneuvers [where live weapons with real ammunition are used]. . . . The peer leadership is tough. . . . Just the experience of leading your peers, 'cause if you can lead your peers you can definitely lead your Marines. God what are some other experiences.?

> LIEUTENANT ADAMS: Participating in raids rather than just RSO-ing [observing][1] them and seeing it.

> LIEUTENANT NANCE: Exactly, yeah. Understanding what these guys are going through, what your Marines are going through. Not just being the one on the hill, like he said, overwatching the operation.

These lieutenants' comments reveal a very significant feature of developing Marine leadership at TBS: the importance of hands-on experience in building credibility and confidence. Applied learning is a primary component of TBS, and Marine officers are required to undergo training in all of the skills that their enlisted Marines will learn at the recruit depots. Tactical decision games (TDGs) also form a significant part of the curriculum. These games place the lieutenants in different scenarios and require them to solve the problem, usually as a team. Often the games are rehearsed through the use of sand tables— a structure that resembles a very large sandbox populated with symbols that can be moved to demonstrate activities or events. "Learning in the classroom is important, but experience is everything. Theory in the classroom, we learn 10 percent; role play at TDDGs [Tactical Decision Discussion Groups] maybe 20 percent. The execution, that's where we learn 60–70 percent," explained Second Lieutenant Kaplan, a student at TBS.

This emphasis on experiential learning is closely connected to the Marine Corps' willingness to experiment and make mistakes. "One of the things I learned as a second lieutenant. There is something called zero based defect [no mistakes]," began Lieutenant Colonel Kramer, a colleague of mine at the university. "But you can't have a 0 defect mentality—it doesn't allow people to practice, to learn things. We understand things won't come out 100 percent. The solution may be 60 percent. The idea is that you can't wait until everything is perfect." Indeed, rather than trying to get something perfect, Marines are encouraged to aim for "the 80 percent solution."

The value of learning from one's mistakes and being willing to try new things was emphasized by Lieutenant Adams, who offered some advice for those following him at TBS: "Give it a shot. Here's where you learn. This is where you make your mistakes. Just because you make mistakes and screw up completely doesn't mean that you are a bad leader. It just means that you tried something. It didn't work. I never learned anything from all my successes, what I learned was from all my mistakes. TBS will teach you how to carry the weapon, how to operate amongst your peers. But if you don't want to put yourself out there and fail, you have no business being here. And that's what I would say."

This view of the leader, not as infallible but as someone who is willing to solve the problem and fail occasionally, is not necessarily typical of most militaries. Indeed, the Marine Corps is unique among most militaries around the world in two other key respects: its view of leadership as an obligation to those being led and its emphasis on encouraging decision making and leadership among junior members of the Corps.

"Leading from the Front": The Obligation of Officers to Their Junior Marines

In many military services around the world, the role of a leader and decision maker is confined to a few senior commissioned officers who view their exclusive position of authority as coming with a variety of privileges: better housing quarters, better food, access to the best gear or clothing, more relaxed requirements for physical fitness, and/or the opportunity to hand over unpleasant and demeaning tasks to someone inferior in status. In contrast, in the Marine Corps, increased rank and leadership are not viewed as coming with extra privileges, but with the extra obligation to care for those being led. Equally important, rather than the centralization of decision making and leadership found in many international (and some U.S.) militaries, Marines speak of "pushing down" leadership and decision making to the lowest ranks. It is expected that all Marines (both commissioned officers and enlisted Marines who are promoted above the rank of lance corporal) will hold positions of responsibility and become leaders. As a result, leadership is not a quality restricted to a privileged few at the top, but an ongoing relationship between senior and junior officers and between enlisted and commissioned Marines.

Perhaps one of the most unusual aspects of Marine Corps culture is the view that good leaders put their junior Marines first. Marine officers emphasize that they are the last ones to eat. They are the last to go to sleep at night. And in a shortage, they will forgo critical clothing or gear in order to make sure their Marines have what they need. According to Colonel Cole, "We [leaders] focus on two things: 1. Accomplish the mission, and 2. Take care of your Marines." Colonel Simons describes leadership as an obligation and responsibility. "It's all about what the leader owes the led." To illustrate what he meant, he told me the story of "salad sandwiches":

> In the field, Marine officers always eat last. During an operation in the field we try and provide one hot meal a day to Marines. Normally the food is brought forward from the field kitchen in "VAT cans" [insulated thermos-style five-gallon metal containers]. The cans are set out on the ground and Marines will file by and get food from the VAT cans—the most junior Marine in the group goes first. I couldn't tell you how many times I ate "salad sandwiches" in the field with my Marines–because the only thing left was bread and leftover lettuce.

To illustrate the unique role of leadership in the Corps in comparison to other U.S. military services, Colonel Moore described the procedure for unloading a ship while his Marines were sailing with the Navy on a MEU (Marine Expeditionary Unit ; see chapter 3). Among the Navy, the captain[2] (the most senior naval officer) of the ship disembarked first, followed by the officers, leaving the lowest-ranking sailors to stay behind and do all the cleaning up and unloading of ship gear. In contrast, when the Marines disembarked, all the work was completed before anyone left the ship. Then the most junior Marines exited the ship first, followed by the junior officers. The most senior officer did not disembark until all Marines had left, remaining responsible for all of his or her Marines until the work was done.

Marine officers speak of "leading from the front." As mentioned above, officers expect themselves to be as capable, if not more capable, than their junior Marines. This includes being equally as fit as their junior Marines, and even leading the unit's running or PT (physical training) exercises. Poignantly illustrating this ideal, at the recruit depot on Parris Island, the final run around the base prior to the recruits' graduation is led by the depot's commanding officer—a two-star general—or his or her next in command. Given that the recruits are generally eighteen-to-twenty-two-year-olds who have just finished twelve weeks of serious training—while the general is probably in his or her fifties or even sixties—that is quite an expectation!

"Leadership from the front" is not simply based on physical fitness, however. Marine leaders are expected to be exemplary in all aspects of their life—physical, mental, and moral. As Captain Hutchinson noted to me, what is important is "leadership by example. A good leader puts his Marines before himself." This notion that a Marine officer should sacrifice himself for his junior Marines echoes strongly of the Spartan ethic I described in chapter 3. It is also strongly linked to the Marine Corps ideal of "every Marine a rifleman" discussed in chapter 2—the belief that no Marine is more important or more necessary than another. Marines' infantry or ground-centric view of the world combined with their strong warrior ethos also explains why a Marine officer would view his or her junior Marines who are out there fighting the fight as more important than his or her own work as leader. Indeed, Lieutenant Nance takes this concept one step further, describing his enlisted Marines as the "real Marines":

> Marine officers, we command Marines. That's what's special about being a Marine officer. But it's a different thing to take command of Marines because, I mean yeah, we never ask our Marines to do something that we wouldn't do. But typically we don't do what we ask them to do. I mean, that's by virtue of our billet. And it's a different beast. We train Marines different than we train Marine officers. We refer to them as Marines, but we refer to officers as officers, Marine officers, because we hold an office. We are a Marine that holds an office. They're a Marine period. End of story. That's it. In some ways we are hyphenated Marines—we command Marines. I think that's pretty common how most people say, "The Marines need this." And they typically aren't referring to Marine officers. They're typically referring to a Marine wearing chevrons with crossed rifles [the symbol on the collar indicating an enlisted Marine].

Not only are enlisted Marines seen as the ones out there doing "real Marine stuff," but the Marine Corps also expects its enlisted Marines to play important roles as leaders as they move up the ranks. Colonel Simons describes this transformation from enlisted Marine to a leader: "An enlisted Marine from the get-go, for the first few years their focus is being led. They'll tell you at boot camp, it's unhesitating obedience to orders. That's why they spend so much time at drill." He continues, "At the enlisted side at about three years we have to start making that shift to being a leader. The hardest shift that any Marine will make is from lance corporal to corporal." The colonel then added a most interesting statement, reinforcing the belief that no Marine is more important than any other: "There's not a lot of difference between a colonel and a corporal. It's kind of span of control—level and scope. The biggest difference between a junior Marine and a junior lieutenant? The lieutenant is a leader for his whole life."

Enlisted Marines who are promoted to positions of leadership are referred to as non-commissioned officers, or NCOs. In Marine Corps military structure, starting at the level of the platoon (which is composed of three to five squads, or approximately fifty Marines—see Appendix), all commanding officers are paired with an NCO. This partnership places the commanding officer and NCO as a team, sharing the burdens of leadership. The important role of the NCO continues throughout the Marine Corps hierarchy. Indeed, even the commandant of the Marine Corps is paired with the sergeant major of the Marine Corps. Frequently, in any major announcement to the Corps, both will speak as a team to the Marines, reflecting their uniquely paired positions and partnership.

Interestingly, emphasizing the role of NCOs as respected leaders, officer candidates at OCS are trained by NCO drill instructors. "Officers are purposefully kept in the background," Colonel Simons points out. "The recruit will primarily see the sergeants and the staff sergeants. We want them to see the NCOs as their leaders. It teaches them to value the role of the NCOs. Only a few militaries do that—those associated with the Brits. The Brits were the first to build a professional cadre of NCOs."

Leadership is such an important value in the Marine Corps that, in contrast to many other military services around the world, even the most junior Marines are encouraged to take responsibility and make decisions. "In the Marine Corps we empower junior leaders," Major Bennett explained to me. Not only is leadership expected of junior officers, but even junior enlisted Marines such as corporals are expected to make their own decisions, some of which may have strategic impact.[3] "The strategic corporal. We rely on that junior leadership," began Lieutenant Colonel Mason during a class on Marine Corps culture to a group of foreign military officers. "Every action that they do will have some strategic impact. We are OK with pushing that corporal out far. We have trained them to be able to do that."

Although there is probably not a military or civilian organization around the world that does not value leadership, as I argue in this chapter, the particular nature of leadership and the ideal qualities of a leader are unique to the culture of the organization. In the case of the Marine Corps, leaders are valued for their ability to be decisive, to take action, and to be willing to make mistakes in order to solve the problem—all traits that are essential for working in the kinds of rapidly changing environments that Marines face as the nation's emergency response team.

Equally important, and clearly tied to their Spartan identity and their focus on "the guy on the ground," for Marines, leadership is not seen as an opportunity to gain personal perks, but rather as an obligation to put the team first and set one's personal needs aside in order to "care for one's Marines." Finally, given the classic hierarchical nature of most military organizations, one of the unique aspects of Marine Corps leadership is the Corps' expectation that leadership and decision making should begin at the lowest ranks. This almost egalitarian ethos reflects the fundamental belief that at the core, all Marines are interchangeable: "every Marine is a rifleman." As Captain Rhodes states, "I think that all Marines are the same. Officers are the same. NCOs are the same. All that changes is perspective over time."

11 REALITIES: "MARINIZING" CULTURE

DRESSED IN A FLIGHT SUIT, PRESIDENT GEORGE W. Bush addressed the American public in a televised victory speech from the flight deck of the USS *Abraham Lincoln*, stationed off the coast of San Diego, California. "Major combat operations in Iraq have ended,"[1] he stated on May 1, 2003, less than six weeks after the initial invasion of Iraq. For a brief moment, it seemed that, as with the Gulf War (1990–1991), the U.S. military and coalition forces had defeated Iraq in a rapid and resounding victory. With their superior technological power and some of the finest and best-trained militaries in the world, the coalition forces had found no meaningful resistance from the Iraqi military. As soon as a new Iraqi government could be put in place to replace the deposed Ba'athist regime, the troops would go home to their families.

The quick return home never happened.

The first signs of trouble were a series of relatively peaceful protests. Then the attacks on coalition troops began. On July 7, 2003, BBC News reported that two American soldiers had been killed in separate attacks by Iraqi civilians—one with a homemade explosive device, or IED.[2] By August 19, the hostilities had spread beyond attacks on military personnel: nineteen United Nations special representatives to Iraq were killed in a bomb explosion at their headquarters in Baghdad.[3] Then on November 8, the Red Cross, which had operated continuously in Iraq since 1980, withstanding three terrible wars there, announced it was pulling out of Baghdad and Basra out of "concerns for staff safety."[4]

Something had changed. Operation Iraqi Freedom (OIF) was turning into a conflict completely unlike (or so it seemed) the clearly defined heroic wars and the famous battles that Marines remembered so proudly in the Marine Corps museum: the Battle of Belleau Wood in World War I, the raising of the flag on Mount Suribachi after a bloody fight to seize Iwo Jima during World War II, and even more recently in 1991, Operation Desert Storm in Iraq. In contrast to the clearly defined lines—blue for friendly forces, red for enemy forces—Marines and their service and coalition partners were suddenly faced with an Iraqi population that any moment could be friendly, hostile, or both. The situation had become, as Rupert Smith famously stated, "wars *amongst* the people."[5]

In military circles, new terms for this strangely unfamiliar kind of war began circulating: "irregular warfare," "hybrid warfare," "complex warfare," "small wars"—all apparently not really war, but some kind of modified, hyphenated thing that the United States and its Western military partners had not been trained to fight. Curiously, however, despite the remembered history in the Marine Corps museum that reinforced the fierce Spartan warrior identity of Marines, in fact, at least for the Marine Corps, this new "irregular warfare" was not new or probably irregular.

Due to its expeditionary nature, the Marine Corps has certainly had some of the longest and most in-depth experience of any U.S. service in conducting small wars and interacting closely with less-than-friendly foreign populations.[6] For over 200 years, Marines have undertaken countless small-scale missions in remote places around the globe. From the Barbary Coast to the Banana Wars to the Combined Action Platoons (CAPs) in Vietnam, Marines have learned foreign languages, operated among and alongside local people, and trained and deployed with foreign militaries.[7] During the Banana Wars in the early 1900s, for example, many Marines became fluent in Spanish in order to understand the local people in Guatemala, Haiti, and Nicaragua. As a result of its experience during the Banana Wars, the Marine Corps produced the *Small Wars Manual*.[8] This guide continues to be one of the most important U.S. military references on small wars today. And although the Marines (and most U.S. services) do not devote much time in their history lessons to the Vietnam War,[9] in fact, Marines experienced more than modest success in fighting another large U.S. "hybrid war" through their CAPs, embedding platoons of ten to twelve Marines in villages in South Vietnam.[10]

And yet choosing to staunchly remember and train for a more heroic warrior history of "conventional wars," Marines—like their service and coalition partners, the U.S. Department of Defense, the U.S. Congress, and even the American public—found themselves initially frustrated by the strange supposedly new "irregular war" that they faced in Iraq in 2003. Part II of this book examines what happened to the "traditional" Marine Corps warrior ideals when the contemporary realities of conflict no longer matched with that identity. How does the ideal tough Spartan Marine suddenly start to solve problems by drinking tea with the sheikh instead of shooting him? The problem from the Marine point of view, as Master Sergeant Wright stated, was that "you can't kick in the door one day and then go in expecting to play Mother Theresa the next."

However, this identity shift is exactly what the Department of Defense, U.S. Congress, and senior leadership of the Corps have demanded of Marines during the past seven years.

DoD Culture Policy, the Marine Corps, and the Struggle for Identity

In January 2005, in response to varying external pressures by the U.S. government, coupled with the U.S. military's realization that its lack of language and cultural skills was hindering progress in the war in Iraq, the Department of Defense (DoD) released the landmark "Defense Language Transformation Roadmap."[11] This document required all U.S. services to significantly improve their capabilities in language and regional area skills. During that same year, the Defense Language Office (DLO) was also given a mandate by the U.S. Congress to develop language policy and oversee the implementation of language programs in the U.S. military services.

One month later, General James Mattis (at the time Lieutenant General in command of the Marine Corps Combat Development Command), directed that the Marine Corps establish a formal training school and Center of Excellence—the Center for Advanced Operational Culture Learning (CAOCL)—to provide urgently needed basic culture and language training to the troops. Shortly thereafter, most of the other U.S. military services established their own respective culture and language programs and centers. Confusingly, each of the service's programs had its own (often extremely loose) guidance from its own service leadership. As a result, very quickly, policies and interpretations of what "cultural training" and "language skills" meant began to vary significantly among the services. The greatest discrepancy, however, was not between the services but between DLO—a language-focused office—and the various services. Given that Congress had mandated the Defense Language Office to develop policy for *language* programs, DLO viewed language as the primary focus of its efforts; the services, on the other hand, quickly began to take another direction.

By June 2007, recognizing the need for departmental integration, the Department of Defense held a summit on language and regional and cultural expertise, and subsequently released a document requiring that the department develop an overarching cultural and language strategic plan.[12] Subsequently, the Defense Language Office stepped in as the "primary lead" for the implementation of this plan. A rather interesting dialogue and negotiation began between the individual military services and DLO—an event worthy of study in itself, but too complex to be narrated here. In the process, the policy debate reached up to the U.S. Congress, which sought to provide additional guidance to DoD, DLO, and the various military services.

Then, in November 2008, the House Armed Services Committee (HASC) released its own study on DoD's progress in "building language skills and cultural competencies in the military."[13] Although HASC acknowledged that the services had made some efforts to develop cultural and language strategies and to implement training and education programs, the report found that in general, the services still had a long way to go. In particular, the report noted that "we are left with several important questions. For example, the Department set a goal of creating foundational language and cultural skills in the force. *Yet the Services' primary efforts appear to be far more aimed at developing a culturally aware force than a linguistically capable one*"[14] (emphasis my own).

Indeed, HASC was correct. By this time, most of the services had begun focusing heavily on cultural skills, and much less on language skills—a shift that I describe in detail in the following chapters. Realizing that perhaps language and culture were separate concepts and not synonyms, in 2008 the Defense Language Office created two positions (out of its forty-two staff members) that would oversee the culture portion of the policy implementation. In contrast to the rest of DLO, this miniscule, culture-focused group (it slowly expanded to six personnel and then gradually reduced again) took a different approach to the policy implementation process, seeking to learn from the various services, in an iterative, shared policy process.

According to Dr. Greene-Sands, who assumed one of the two initial DLO culture policy positions, "With regard to language policy, the services were given an 80 percent solution before requesting input, which often caused pushback or non-concurs [dissenting votes] on policy. But Brad, [the director of the DLO culture group] said, 'Before we write any [culture] policy we will get a pulse check and see what the services would say.' . . . That's when I started to ask for input. Basically [when we arrived] everyone in the culture centers had already started moving out on this. So the cart was before the horse already. The culture centers were already in place. Instead of leading, we started alongside. [We've focused on] building consensus across the services—that's huge."

Reflecting this effort to integrate various service perspectives and approaches, DoD then convened a series of Regional and Cultural Capabilities Assessment Working Groups (RACCA-WG) composed of representatives from across the services. The purpose of the working groups was to develop a common terminology and shared set of standards regarding cultural and linguistic competencies. Although the committee made a number of recommendations, none, however, were formally adopted by any of the services.

In the ensuing years, the Defense Language Office has slowly but persistently sought to create a coherent policy on cultural and language learning and skills for the various military services. Responding to feedback from the various U.S. military services, cultural capabilities have been identified separately from language skills and collected into a generally accepted concept called "Cross-Cultural Competency," or "3C" for short.[15] In 2011, the Department of Defense issued yet another culture and language document—this time the "Strategic Plan for Language Skills, Regional Expertise, and Cultural Capabilities."[16] The plan set out three goals: "identify, validate and prioritize language skills, regional expertise and cultural capabilities"; "build . . . a Total Force with a mix of language skills, regional expertise and cultural capabilities"; and strengthen these capabilities.[17] Two years later, in 2013, another policy document was released describing a "measurable" set of language, regional, and cultural capabilities that commanders could use to identify experts for staffing purposes.[18] Recognizing that cultural—as well as language and regional skills—are valuable and necessary for operating among foreign peoples, the new policy, titled "Language, Regional Expertise and Cultural Capability (LREC) Identification, Planning and Sourcing, CJCSI 312.01A," specifically identifies language and culture as separate skills.

Yet although these documents appear to validate the various services' experiences that culture is not synonymous with language, it is clear by reading the policies that language, not culture, continues to be the priority skill for DoD. For example, the 2013 LREC publication devotes over 60 percent of the document to discussions of various language skills and requirements, while culture receives a scant six-to-eight-page discussion in the sixty-three-page policy.

While the Department of Defense and the Defense Language Office were busy trying to develop a consistent culture and language policy for all the services, the U.S. Congress was also busy trying to determine if, in fact, the various services were actually making progress in adapting to the various cultural imperatives and policies. In 2011, Congress directed the Government Accountability Office (GAO) to evaluate the extent to which the Army and Marine Corps' programs, training, and strategies were aligned with DoD language and culture planning efforts. The resulting report[19] found that while the Army and Marine Corps had developed culture and language programs and strategies, the programs were inconsistent, and there was no synchronization between the services. Furthermore, neither service had adequate statistical data for GAO to evaluate the effectiveness of training or outcomes ("metrics").

Interestingly, reflecting the commonly accepted notion that failures in policy implementation must be the result of either poor or weak leadership and/or unclear policy language and guidance, GAO concluded that "DOD has not yet set up internal mechanisms, such as procedures and milestones, by which it can reach consensus with the military services on priorities and investments."[20]

As I illustrate in the following chapters, however, the challenges of implementing culture and language policy are not simply the result of poor or inconsistent policy guidance from the top. Rather, as the stories of the Marines in my study reveal, Congress, DoD, DLO, and the Marine Corps have each held different ideas of what they need to do to solve the problem of successful interaction with foreign populations during war. These differing perspectives have led to contrasting expectations of what "successful implementation" looks like. In fact, the issue extends beyond different ways of framing the problem between the services, DoD, and Congress. As I illustrate in chapter 7, civilians' (and particularly the subject matter experts who work with the military and create policy) and Marines' worldviews are based on very different assumptions of what culture is, how it is related to combat, and how and why cultural (and to a lesser degree language) skills can be applied to effectively resolve the conflicts in which Marines are involved.

Part II of this book focuses on how the culture of the Corps has influenced its responses to and interpretation of both the "irregular warfare environment" of Iraq and Afghanistan as well as the ensuing policy directives. The result may appear (to outsider government civilians and scholars) to be a rather convoluted and nonlinear set of solutions. Yet for Marines, their amalgamation of new external realities with their own traditional ways of conducting business reflects a series of creative compromises and solutions that fit within Marine Corps processes and ways of seeing the world. In the case of the Marine Corps, as I illustrate in the following chapters, adaptation to these external pressures has occurred through four overlapping approaches: *simplification* (chapter 6), *translation* (chapter 7), *processing* (chapter 8), and *reshaping* (chapter 9).

6 "Building the Plane as We're Flying It"

Simplifying Solutions to Culture in Theater

Sayyid's office was decorated with numerous plaques from Marine Corps units thanking him for his service: photos of him standing next to general officers and other senior military officials in Iraq, an Arabic inscription with verses from the Qur'an and decorations from the Middle East. "I did active duty in '87–'92 and then I got out from the military," began Sayyid, an Arab American who had been born in Jordan. "I came back after September 11 of 2001—as an active reservist. A lieutenant colonel called me at my home. I'd been out of the military for ten years by then. I don't know how he even got my phone number! And he asked, 'Would you be willing to help?' I know how the Arabs feel when a foreigner comes to their country. So I thought, 'I could help on both sides.'

"On a Thursday, out of the blue, I got an email—I had orders to travel to Camp LeJeune on Sunday and continue on to Kuwait. I was actually a hotel manager at the time, earning six figures and then here I was, [earning a quarter of my salary as] a sergeant—which is what I was when I got out. So I was activated as a sergeant.

"I landed in Kuwait. They picked us up. The camp was in the airport—a thousand of us there in cots. I had no clue who I am reporting to or what I'm going to do. At 10 o'clock at night I'm sitting there thinking, 'WHAT am I doing here?' when some guy walks in—Chief Warrant Officer Jones, a Marine—and shouts out, S-A-Y-Y-, starting to spell my name. And I thought, 'That's me!'"

"How many duffel bags do you have?'"

"Four."

"Well by 0400 you're going to get down to one."

"I was pissed because there was nothing explained. At midnight I demanded to see a general officer. Of course I didn't get him. But I what I landed was a lieutenant colonel. Finally what it came down with is that they are short of interpreters and I am the only one who speaks Arabic and knows the culture. . . . We ended up landing in M——. Buildings went up overnight. I was sitting there in a tent and I run into this civilian interpreter that was from Iraq—but an American. [He tells me] I've landed in a Force Recon unit. But I didn't know what it is. So the interpreter said, 'You know what Force Recon is? It's like Special Forces.' And I'm thinking, 'Special Forces! I have three small daughters.'

"[So I go to the Force Recon.] These guys are looking at me and they're large and fit and here I am an old man with a belly. But when I got in they jumped in and were all over me. They were dying to know about the religion, the language and culture. I was lucky to land on a smart unit. . . . Then they took me to the enlisted guys. Later the enlisted guys told me, that [when they saw me the first thing they thought is], 'Oh my God, here's an old guy—he's Arab and Muslim. What have we got?'

"When they got their orders that our mission was Intel [intelligence], they shifted their focus from fires [kinetically focused actions using such methods as mortars, artillery, air power]. Almost immediately I went directly from the mode of being an interpreter to the mode of being an advisor to them. . . . Our mission was to go in to the police there, the tribal leaders. I did not sleep. We'd conduct a raid, sleep an hour or two, and then go out again to talk to leaders.

"Most of what I focused on was gathering info, working with the Iraqis. Some [U.S. military] units were out there doing some pretty bad stuff. That was difficult and I would go out into the town and I would hear the stories. And that's where the people started fighting us. They were normal people like us, but now they wanted revenge.

"Things escalated really quick. We started doing a lot of raids. But 70 percent of the raids were [done] wrong. Marines, when they conduct raids, it was a pretty barbaric type because that's what we were trained to do. So I was the guy who defused the situation. I was the guy at the front at the door. I approached cautious, but I didn't have the intent of 'I want to kill people.' Marines were going in saying, 'We're going to clear all the people.' I got them away from that. I would look through the paperwork because I could read Arabic. So Sayyid's billet went from interpreter to counterintelligence agent to interrogator. . . .

"There was a lot of situations, many stories like that. Some guy is running to his brother who was shot and the Marine wouldn't understand him and shoot him. . . . The cultural knowledge and advice that I gave to these guys, I think that was very critical. The first assault, they went in and they were pulled out and they were pissed. They didn't want to play politics, but eventually they learned. But I think the Marines, particularly General Mattis, were influenced and thought that particularly the culture was important. The Marines learned immediately and covered the need for this in '04. They didn't know it [the importance of culture] in '03. But they did in '04.

"My news reached everybody—reached Toolan, reached Mattis. On my departure, they had a formation for me. It was very emotional. The Force Recon is very proud of themselves. And they said, 'We could never have had as good a Force Recon as with you.'"

WITHIN A FEW MONTHS AFTER THE INVASION OF Iraq, Marines rapidly began to realize that to be successful, they would need to understand and work with the people there. Reflecting the Marine Corps bias for action, their focus on mission accomplishment, and their willingness to be flexible and adaptable in order to succeed, Marines initially took the lead among the U.S. services in finding quick and simple solutions to the cultural problems they faced. Far ahead of subsequent congressional mandates and Department of Defense (DoD) policy directives, by 2004 Marines were already developing innovative approaches to the new human aspects of combat, proudly at the "tip of the spear" of the problem. As Sayyid's story illustrates, they turned to their translators to provide not only language interpretation but also cultural analyses and even cultural solutions. Realizing that overly aggressive interactions with the civilians in the country were counterproductive to their mission of creating stability and peace, a number of Marine units like Sayyid's Force Recon group adjusted their way of operating, relying on translators and other cultural experts to smooth interactions. In fact, the Marines' and other U.S services' successes in the "Anbar Awakening" in Iraq have been attributed to this early willingness to adapt to the cultural aspects of the conflict.[1]

However, the Marines' rapid, ad hoc, and improvised solutions also resulted in a number of both short- and long-term challenges. Many Marines I interviewed saw cultural adaptation as simply another form of "semper gumby"—just something that Marines had to do to get the job done. Yet by viewing cultural problems as just the latest operational challenge, Marines assumed that the issues they faced in interacting with the population were limited to the current conflict in Iraq. *Simplifying* the problem, Marines sought quick, short-term, easy solutions that would require little to no long-term change in Marine Corps ways of seeing the world or conducting wars.

As a result, initial cultural challenges in interacting with the population became simplified and defined in three ways: as something that Marines do anyway; as a decoding challenge requiring translators and language training; and as an ethical problem.

"Semper Gumby": Culture as Just Another Marine Way to Fix the Problem

> "Marines understood culture even if they didn't understand it, if that makes sense. They knew that it was important. The majority get along with people, understand people."
> —*Sayyid, Arabic interpreter, speaking of working with Marines in Iraq*

In my conversations with Marines about their early experiences and challenges in Iraq, perhaps one of their most common responses was that cross-cultural interaction really wasn't all that hard. In their view, any Marine with a basic understanding of human nature could quickly figure this out. Because Marines saw themselves as flexible, adaptable, and willing to work out innovative solutions to their problems, many viewed dealing with cross-cultural challenges just like any other problem they faced. Coping with difficult, unexpected, messy situations (including cultural differences) was just par for the course, not a catastrophe that would hinder Marines.

"If you think you know what you're going to be doing, you don't," Major Bennett stated to me as we discussed his recent deployment to Iraq.

"I was a member of a one-year MTT [military transition team]," Major Bennett began the interview, sitting tall in his tan camis (camouflage uniform) in my office in the winter of 2007. "When we got in theater . . . the Iraqi [army] unit had been formed. They had been piecing soldiers together in August '05. The momentum was to get the unit ready before the elections. When we got there the camp was set up; they were sort of getting paid, sort of getting food. It had a fledgling headquarters, it was young—with only 40 percent of the Table of Organization [staff]. Some equipment—not all they needed to perform counterinsurgency operations."

"We were a ten-man team—a lieutenant colonel as a team chief," he continued. "The next level we had three majors. . . . We had one other officer who was basically my assistant operations officer who—I don't think it was really designed that way—but he really became more my tactics instructor. We kind of morphed roles to our strengths. And then the remainder of our team was five enlisted Marines with varying specialties." He paused to collect his thoughts. "While we were over there, I think all MTTs evolve—or devolve—that's a good word. We would add attachments from the RCT [Regional Combat Team]. Unfortunately, our MTT team sustained 30 percent casualties throughout the year. The core was always there, but to think that the ten-man team stayed together for the whole time. Not close.

"Like any embryonic program like the MTT program, there were holes in the program itself. But the essential tools were provided. And as a Marine you did the things that are expected from Marines—which means be creative, be flexible, be resourceful. Then there's no reason why you couldn't succeed. Our success was not an independent Iraqi battlespace, but we did achieve Iraqi-led operations."

Major Madison, who had also deployed on a MTT in Iraq, expressed similar ideals that Marines were capable of handling whatever was thrown at them and solving the problem with the limited resources they had available. During his deployment, while he did not succeed in his stated task (intelligence training), he did develop innovative, culturally appropriate solutions that assisted the local population and built goodwill between the Marines and the Iraqis.

"When we came I thought I was going to be teaching about the Intel [intelligence] cycle and it never occurred. . . . By billet I was the Intel trainer. In reality, I got on deck [arrived] and it was like, 'So who's your pay guy? Madison you're it.'

"[The job] was far from making sure direct deposits kicked in. . . . Here we were in Anbar. I would fly with three Iraqi pay officers into the Green Zone in Baghdad. The minister for defense would cut a check for the brigade—pay $500 per soldier. I would personally drive an armored SUV, go to the local bank, and carry out six potato sacks of cash into the SUV. I'd fly with a million dollars worth of cash, about a billion dinars. Then we'd have to convoy to the battalions, drive about 150 miles to different battalions. Then fly back to Baghdad, return the extra cash and reconcile the pay rosters.

"We had ghost soldiers—names on the pay rosters where there was no such person. On the other hand, we had 300 soldiers who were not on the payroll. We had soldiers serving and who had even died that were never paid. So my task for about a month, two months, was a daily fight to get rid of ghost soldiers and paying unpaid soldiers.

"There was no local bank in Anbar that supported the [Iraqi] military. Our team was able to establish the Arab Bank. Then my process was flying to Baghdad, picking up the check, and bringing it to the bank. We brought economic advantages to the community by showing our good faith."

"We're building the plane as we're flying it" is a picturesque Marine saying that captures this notion of inventing and creating solutions while in the midst of a crisis situation. This Marine can-do, "fix anything" attitude enabled many of the Marines I interviewed to respond quickly to the initial cultural challenges they faced in theater.

"I think we had to change more than the Iraqi army did," Major Bennett noted. "The first couple of months we . . . kind of pulled back the reins on them to make them what we wanted, and not what they were. Then over the course of time, I don't want to say we lowered our standards, but there are things that Marines do differently than the Iraqi army would. Rather than spending nine months making them Marines, we just thought we'd try to make them into a COIN [counterinsurgency] capable unit. At least for me personally it was me changing as much as the folks we worked with."

Noting that the main problems were not necessarily cultural understanding but working with his own inflexible U.S. military system, he added, "I think back to the skills we were trying to teach. It was a reflection that some American units couldn't break our conventional DNA. Pretty soon we kind of threw the playbook away. It was an inappropriate playbook. It didn't do anything for the skills we needed to do." Reiterating the Marine value of being small, Spartan, and agile, he added, "There are certainly advantages to being a MTT team. We were small and as long as we had ten guys on the same sheet of music [working together] and we were not violating the UCMJ [Uniform Code of Military Justice], we had the flexibility to do it. . . . To a certain degree we were able to do what Marines like to do: give me some endstate, some guidance, and a roadmap and we'll get it done."

Similarly, although Major Bennett was fully aware of the cultural differences between the Marines and the Iraqi army he was training, he saw these differences as being easy to overcome. With a bit of imagination, creativity, and willingness to adjust, even language barriers were not necessarily an issue:

> [Although] it was a challenge dancing around what it means to be a Muslim, we were kind of well tuned into all that before we left—at least we'd heard, "be aware of it." But how to handle Fridays; how to handle Sundays 'cuz that's kind of our day. How to handle Ramadan. I guess there would be instances where there was kind of a curve ball that was thrown your way. . . . Some of those are challenges but there was almost too much buildup that we had received about cultural differences. We knew they were different and they know we are different. . . . In fact, language—at least with a couple of them—was never an issue. I carried a dry erase board, and we both talked through military symbols. So you don't have to know the word to say mortar, you just draw a mortar symbol.

Major Madison expressed a similar view that adapting to the Iraqi culture and advising was not difficult; the harder part in some ways was getting the Americans to understand what they were doing, "You didn't need to be a CSC [Command and Staff College] grad to be an advisor. You've got to have solid leadership skills—not only for negotiating with Iraqis, but with our own American counterparts. They wouldn't understand anything of what we do."

This belief that cultural skills were a necessary—and expected—part of what Marines needed to do to succeed in theater was not simply restricted to Marines on military transition teams, however. In the Center for Advanced Operational Culture Learning (CAOCL) survey, when asked to rank the importance of culture and language to military operations, Marines (both enlisted and officers) ranked culture on average as 3.25 (somewhere between important and very important) and language as 3.09 (important).[2] As Table 6.1 illustrates, Marines in the survey ranked cultural skills highly in terms of their importance and relevance to military operations.

TABLE 6.1 Marines' attitudes regarding the importance of cultural skills

Attitude questions	Scale	N	Mean	SD
22. Marines need multicultural skills to be operationally effective.	1-Strongly disagree 5-Strongly agree	2,377	4.00	.785
23. Based on your experience in the Marine Corps, how important to mission effectiveness is understanding the local culture?	1-Not important 4-Very important	2,385	3.25	.726
24. Based on your experience in the Marine Corps, how important is culture training compared to all the other pre-deployment training requirements?	1-Less important 3-More important	2,385	1.97	.528
27. Having the skills to understand the impact of culture on any given operation is a key component of operational readiness.	1-Strongly disagree 5-Strongly agree	2,365	4.07	.718
28. Having Marines in your unit knowledgeable of different regions of the world is an important component to operational readiness.	1-Strongly disagree 5-Strongly agree	2,365	4.09	.727

SOURCE: CAOCL survey
NOTE: Using an inter-item correlation matrix for the five attitudinal questions, the Cronbach's Alpha calculated for the five items was .852, indicating a strong positive relationship between all of the attitude questions.

Most interesting, however, were the responses from the Marines who held ground combat specialties (such as infantry, artillery, tank operators, and engineers).[3] As discussed in chapter 2, the "guys on the ground" are considered the heart of the Marine Corps. Although Marine Corps ideology describes the guys on the ground as the hard-fighting, hard-charging "grunts," according to the survey, it appears they are also the ones at the "tip of the spear" in experiencing and responding to cultural challenges in theater. The ground combat Marines reported spending almost twice as much time interacting with the local population (47 percent) in comparison to other respondents (29 percent). For laymen, in statistical terms, the probability (p) that these differences could simply exist by chance is less than 1 in a 1,000 ($p < .000$).[4] Not surprisingly, the "guys on the ground" stated they also used language skills ($p < .000$)[5] and cultural skills ($p < .000$)[6] significantly more often to accomplish the goals of their mission than their combat support partners. And reflecting their view that these skills were important to accomplishing the mission, the "guys on the ground" ranked the importance of using language skills ($p < .000$)[7] and understanding the local culture (p < .000)[8] significantly higher than combat support respondents.

One of the more important results of these survey findings was that Marines considered cultural skills to be significantly (in repeated statistical tests) more important than language for enabling mission effectiveness.[9] One possible reason that emerged from the qualitative data was that in many cases Marines had interpreters who could translate the language for them, and so given the difficulties in learning non-European languages such as Arabic, they preferred to turn to native speakers to assist them.

Culture as a Decoding Problem: Subject Matter Experts, "Terps," and Heritage Speakers

As Sayyid's story at the beginning of this chapter poignantly illustrates, one of the first steps the Marine Corps (and other services) took at the beginning of the invasion in Iraq was to hire or locate individuals within the Marine Corps and other services who understood the culture and language of Iraq. This early solution to the problem reflected the military's and Marine Corps' assumption that culture and language (which were considered as virtually synonymous and interchangeable in the first couple of years of the culture venture) could be "decoded."

One long-established military occupational specialty (MOS) is that of the cryptolinguist—decoders who study the secret messages of the enemy in order to figure out what the enemy is doing. Not surprisingly, then, one initial view of culture was that it was simply another language code that could be broken with the right tools or interpreters. In this view, culture/language was a code that military cryptolinguists or their translators and subject matter experts (SMEs) could easily decode with enough time and resources. The assumption was that the cultural differences between Iraqis and Marines were fairly simple to understand—all Marines needed, actually, were a few interpreters who could help them talk to the "locals."

Thus some of the first outsiders to be employed by the military during the Iraq war were, not unexpectedly, interpreters who not only translated words but also often provided cultural explanation and advice. "Terps," as Marines referred to them, were hired to assist units in speaking to Iraqi leaders and others in the population. Usually these interpreters were heritage speakers from Iraq or other Arabic-speaking countries and were assigned to specific

Marine units during their deployment. As Sayyid observed, "They knew it was very important. They needed to have language and culture there. But they were kind of on the lazy side—they wanted someone else to do the translation, handle the culture for them."

During my research, numerous Marines commented on the importance of their "terps" in teaching them the language and culture of the area. "I cannot forget the interpreters. My interpreters were always teaching us words and ways of the Iraqi people," noted a gunnery sergeant and radio chief in his response to the CAOCL survey. Similarly, a Marine captain and ground supply officer stated on the survey, "Having the services of a local interpreter is invaluable in interacting with locals. Interpreters can understand more than U.S. service members what is actually being said, cultural nuances and exactly what locals are trying to convey based on their cultural understanding."

In addition to hiring interpreters, native-language speakers and other experts were also hired in the United States to provide language and culture training to the troops prior to departure. In part due to the Marine Corps' predilection for quick execution and action, and in part reflecting its inherent flexibility in responding to the mission, initially there was no centralized organization hiring and training these experts. Reflecting the Marine Corps' respect for leadership and decision making at all levels, unit leaders and higher level commanders were given an immense amount of discretion in selecting what predeployment training they conducted, how much time to devote to that training, and who would provide that training. The result was an extraordinary flourishing of many kinds of culture and language training programs, provided by a range of different contracting companies.

Since the U.S. government was reluctant to hire new permanent personnel, in order to meet the sudden and urgent need for interpreters and language instructors, most were hired through private contracting companies (the U.S. government equivalent of temporary employment agencies). As a result, initially Marines were receiving both interpreters and training from many different organizations and individuals, each with their own approaches and ideas as to how culture and language should be taught, interpreted, and understood. Some were contracted by Marine Corps organizations such as CAOCL (see chapter 7), the Marine Corps Intelligence Activity (MCIA), the former Security Cooperation Education Training Center (SCETC), or the Marine Corps Training and Advisory Group (MCTAG). Others were contracted out from DoD organizations (for example, the Defense Language Institute) or other U.S.

military services such as the Army. Yet many other private contracting companies were also providing their own translators, instructors, and materials directly to specific Marine units. And in certain cases, local community colleges or universities were also offering classes taught by their faculty or graduate students.

Since the contracting companies hired the interpreter or instructor, by law it was up to the contractor—not the U.S. government—to screen and train applicants and determine if they were qualified to interpret or teach. This created a number of practical problems in providing Marines with culture and language support in ways that worked for them. First, while a number of these predeployment training instructors were natives from Iraq—and later Afghanistan—due to the difficulty in locating sufficient heritage speakers and culture experts, in the initial years (2003–2008) these instructors came from a great range of backgrounds. Some were Middle Eastern expatriates or their children; others were regional experts who had some familiarity with the Middle East; and a few were simply Marines who had completed a deployment in Iraq and could talk about their experiences.

As several Marines observed in their responses to the CAOCL survey, the hodgepodge of instructors and interpreters was extremely varied in their quality and ability to communicate and teach. A number of them were outstanding. One chief warrant officer, for example, stated, "The best was during a MOUT [military operations on urban terrain] training in 2007 before an OIF [Iraq] deployment where Iraqi nationals lived in the town we trained in aboard [at] Camp LeJeune. They cooked for us, sang songs, and broke us down into small groups where we discussed cultural issues."

However, other instructors and interpreters were not adequately qualified or prepared to work with Marines. On occasion, Marine units would deploy with an interpreter who was not capable of translating. Sometimes the interpreter or instructor did not speak either English or the native language well. In other cases, particularly in areas where many languages are spoken, the interpreter or instructor did not speak the language of the area to which the Marines were deploying. "We deployed to Nuristan province—where no one speaks Pashto or Dari—only Nuristani. Even our interpreters didn't speak the language," noted a logistics officer in his comments on the survey.

"We had a situation where an Afghan SME was supposed to do role playing for a KLE [Key Leader Engagement]. And he didn't speak a word of

Pashtu. So he was speaking Dari and the students were speaking Pashtu and it didn't work," Lieutenant Colonel Jones stated about a negotiation exercise. In another case, one staff sergeant wrote in frustration in his response to the CAOCL survey, "[Our instructor] had not yet mastered the English language and we couldn't understand 50 percent of what he was saying."

Language and verbal interpretation skills, however, were not the only problem. In my interviews and in the CAOCL survey, a number of Marines expressed concerns about the reliability of the cultural information they were receiving or the bias of their teachers. "Instructors need to be screened and teach fact, and not preach their personal academic/moral/religious agenda," commented a Marine captain on the survey. "Academics and expatriates are not necessarily the best or up-to-date. I had multiple instructors talk to me ad nauseam about Afghan culture, Pashtunwali [an Afghan ethical code], do's and don'ts etc. After nine months with the Afghans, I know that these instructors were grossly out of touch with the current state of Afghan 'culture.'" A Marine lieutenant colonel and infantry officer echoed these sentiments when he stated, "Don't rely on culture learning from expatriates who have lived in the U.S. for too many years. They are out of touch with their own people often."

These critiques reflect the importance of up-to-date information and "on the ground" experience for Marines. As described in the first part of this book, according to the Marines I interviewed, credibility is not based on one's degrees or book learning but on one's experience: the person who has seen "the ground truth." Thus reports and instruction from the "guy with boots on the ground" are considered to be more accurate and more relevant than instruction from some SME who has studied or lived in a country many years ago. During a class on leadership at the Enlisted Professional Military Education school, Staff Sergeant Hagerman reflected this emphasis on "real time" information from the field versus academic, book learning: "You need to see what's happening on the ground, go to the most reliable information on the ground, to your unit leaders. You don't go to some manual that's already three years out of date."

The importance of receiving instruction from individuals who have had recent experience in theater was emphasized by one captain who commented in the survey, "I highly recommend finding instructors that have been to Afghanistan in the past ten years, particularly interpreters who have been expatriated and know how to communicate to Marines. And [also hiring] Marines that have deployed to these theaters in recent years."

Even so, some Marines in the survey cautioned that simply having experience in theater was not sufficient for someone to teach about the culture of the country. As one captain wrote in his response to the CAOCL survey, "I felt that the instructor, who was a prior corporal with PTT [Police Transition Team] experience, had no business being a cultural instructor and . . . knew little more than myself."

While the Marines sometimes found their translators and language instructors unable to provide what they thought they needed, on the other side, the native-language speakers had their own challenges. Muhammad, a native Arabic speaker who was hired to teach language and culture to Marines, explained his initial frustration in working with the Corps:

> I saw an ad looking for an Arabic instructor three months after I got here [Muhammad had just moved to the United States]. By chance. The interesting thing was when I came [to the Corps]—the culture shock. It wasn't that big with adjusting to the U.S. But here the first thing that struck me was I realized how I didn't know that much about the Marine Corps. I was really surprised. Here there are Marines and they don't know anything about the Middle East! The second shock was they think they don't need it, they don't care. [They would say to me], "Why do I need to learn Arabic? They all speak English."
>
> From that I started to learn what they need, what they want. I began to learn how to communicate with them because I had to learn their language. Another point that I learned about Marines. They are standing there and they are thinking, "What do you know?" You have to sell yourself. If you don't know the tools you need to communicate in a few words, you either lose them or win them.
>
> I was put with a Marine sergeant to see how he taught culture. He showed me tricks, being a Marine—how he taught the class, how he went around. . . . I learned how to speak—just jumping in, to challenge them without making it sound personal. The sergeant taught me by observing him. It's all about bonding with them. I share with them stories from other Marines and they like it. They see that you are speaking their language and you understand what they need.

As I elaborate in later chapters, this cross-cultural dance between Marines and their SMEs—the difficulty in translating from Marine culture to outside experts they have turned to for help—is a challenge that continues up to the present time. Interestingly, another equally challenging and enduring issue that arose from the early efforts of Marines to respond to the cultural challenges in theater was the ethical problems faced by young, inexperienced Marines.

Culture as an Ethical Problem: Values Based Training

The classroom is bright but austere: ten rows of metal chairs face a wide screen. At the front stands a Marine officer who is in charge of teaching the mandated course on combating trafficking in persons to the forty or so recruits. The students are quiet, orderly, sitting rigidly upright and at attention. The silent formality of the class belies the subject of the course: a lecture on the illegal sale and trade of persons—predominantly child and female prostitution. For the next half an hour, the students and I watch a series of disturbing PowerPoint slides, interviews, and film clips of women and children who had been sold, coerced, and beaten into sexual and other forced labor. Although I had been required to take a similar online class as a civilian employee of the military, I am rather surprised to see this course as part of the core curriculum at the Parris Island recruit depot. Somehow I had not visualized the initial recruit training to include sitting in a classroom on human trafficking, particularly on a clear sunny day when it seemed that practicing drill or running a few laps around the PT fields would be more appropriate.

"If you see these things what are you going to do?" the major nods, then paces back and forth in front of the classroom. Behind her, the screen displays a photo of a white man in what appears to be Thailand, handing money to a toothless man and staring at an Asian girl barely ten years old at his side. "This is not compatible with Marine Corps values."

A hand goes up and a recruit stands up. "What if these things are legal in that country?"

The officer points at the photo emphatically. "The Marine Corps policy is zero tolerance. . . . If you go to some country where it is legal, in the Marine Corps it still is not." The slide changes to a picture of a striptease bar. "When you go somewhere else, to another unit, there will be a list of places you cannot attend. . . . You do not go there. These are off-limits establishments. . . . Marines are subject to the UCMJ [Uniform Code of Military Justice] on and off duty. Seven days a week."

"If people think that we're the elite fighting force and we have Marines doing these things, what will people think of us?" She paces again. "Marines are ambassadors to the world. How we behave is a reflection of the character of the Corps."

Soon after their arrival in Iraq, Marines discovered that the kind of warfare they were facing was much more complex than simply "locating, closing with, and destroying the enemy." Often it was difficult to determine who the enemy was. And increasingly, the solutions to the conflicts in which they were embroiled required an incredible range of activities, ranging from humanitarian aid to fierce fighting. In this "new" reality, warfare seemed not as simple as it was in the past: there were no obvious good and bad guys, which made ethical decision making far more complex.

Compounding the issue, due to modern media the Marines were on stage with the world watching their actions in the news, on television and even on Facebook and Twitter. Given that Marine identity is intimately tied to upholding the Corps' proud and honorable history, as the officer in the trafficking class explained to her students, contemporary Marines must view themselves as ambassadors for the United States. Their actions on and off duty are evaluated twenty-four hours a day, seven days a week.

Unfortunately, as Sayyid pointed out in his interview, while most Marines can and do live up to the ideals of honor, courage, and commitment taught in boot camp, in the early years of combat in Iraq, not all Marines adhered to their code of honor in ways that reflected well on the Corps. The problematic actions, particularly of some of the young Marines who had little experience outside of the United States, not only potentially tarnished the Marine Corps' image but also caused backlash from the local people. As a result, the failure to act appropriately with the local population hampered the efforts of the military to stabilize the country.

Thus while one quick and simple approach to the cultural challenges faced in Iraq was to hire interpreters and obtain culture/language training, a second (and interrelated) way the Corps viewed the situation was to define it as an ethical issue. Again, this was seen as a fairly simple problem: one with which the Marine Corps had dealt periodically over its history. And as such, it had a fairly straightforward solution: ethics training.

During the decade since the initial invasion of Iraq, the Marine Corps has focused on two major training programs aimed at creating more ethically responsible Marines. The first has been a revamping and refocusing of the values based training program (VBT) and Crucible at the recruit depots. The second approach, initiated in 2012, has been a servicewide mandated ethics training program for all Marines.

According to a command brief from the Marine Corps recruit depot at Parris Island, VBT "prepares Marines to make ethical and moral decisions over their careers and during their lifetime."[10] The Marine Corps has not always had a values based training program, however. In fact, its recent history has demonstrated a rather uneven commitment to the concept. In 1996, viewing the ethical aspects of the Marine as a critical formative part of recruit training, General Krulak (31st Commandant of the Marine Corps) added one final culminating test for the recruit depots to prove that the recruit has developed not only the physical but also the mental and moral qualities required to become a Marine: the Crucible.

While the Crucible has been a part of recruit training for the past fifteen years, its role particularly in teaching ethics has shifted several times. During the past two decades, the Marine Corps has struggled with the question of whether it should focus on core values and ethical decision making—a more "human-centered approach" to combat—or emphasize more physical basic warrior skills such as rifle training and combat training. This fascinating internal debate regarding the Marine Corps' identity has centered in large part around the role of the Crucible. The oscillation between these two approaches was narrated to me by Lieutenant Colonel (ret) Custer, who had lived through it all—first as an enlisted Marine, then as an officer, and now as a civilian developing Marine Corps training programs at Training and Education Command (TECOM):

I went through recruit training in the summer of '91. At that point, under the auspices of General Gray [29th Commandant of the Corps], his efforts were to reintroduce warrior training. General Gray got the Marine Corps thinking about the lessons from Operation Desert Storm. So when I was a recruit, basic warrior and combat training was the focus. Fast-forward to General Krulak and values based training. Honor, courage, and commitment became the big thing, and he instituted the Crucible. After he left—General Krulak was kind of polarizing—either you liked him or you didn't—he left a pretty powerful legacy.

So now it's 9/11 [September 11, 2001]. We've been doing the Crucible for a few years. And we've become a Corps of War. Up to now, the two recruit depots—Parris Island and San Diego—we'd tried to make them identical. Then in the 2001–2 time frame the two depots stopped talking to each other.

There's a big effort to bring back basic warrior training. Both depots start to run their own programs. General Flynn[11] comes down in 2006. Parris Island gives this big brief [presentation] on turning the Crucible into a tactical [physical combat] event like SULE II [small unit leader's exercise] at OCS [Officer Candidates School]. Well General Krulak got wind of this and comes back to visit. Every officer and staff NCO [non-commissioned officer] was required to come into the lyceum, and he gave them the exact same PowerPoint he had given in 1996. The general message is, "Don't mess with my Crucible."

We go back. We have to rewrite the POI [program of instruction] to include values again. We went back and reread General Krulak's articles—the whole millennial discussion. We have these unfortunate events—Haditha, Mahmudiyah—all were pivotal events. And they made us take another look at how we're teaching values. Honor, courage, commitment was becoming a bumper sticker. So we went from General Gray and warrior training, then General Krulak and values based training and the Crucible, then 9/11 and a nation at war. Then some negative events led us to refocus on values based training. A sine wave of tactics versus values versus tactics and now we're back at values again.

The period 2006–2007 turned out to be a pivotal turning point in the Marine Corps' alternation between culture/values training and basic combat skills training at the recruit depots. Both the Haditha killings in 2005[12] and the Mahmudiya rape and killings in 2006[13] focused great negative public scrutiny on the U.S. military services. The result was a decisive shift toward values based training at the depots. Major (ret) Wagner describes this shift:

[At the time] I was a training officer. . . . I had been doing training since 2005. So I was here for the tumult that occurred. . . . In 2007 TECOM sat down and said we have to reset. In November 2006, both the School of Infantry commanders, both depot commanders, and Mr. Judge, who had been hired [for legal counsel], stepped in to get things in order. In 2006, we had to define the training pipeline—the continuum from the depots to the MOS [military skills training] schools. . . . [We had to decide] of the 300 tasks, which are the basic Marine tasks and which are the rifleman tasks. . . . [And then there was] the issue of the Crucible. What is its role? We needed to make it less about enduring pain for fifty-four hours and focus on the ethical dimension.

[Regarding] values based training—a lot of great work was done with General Krulak. But we didn't formalize it. We had a lot of classes on VBT but it was instructor dependent. There was no DI instruction on how to teach it. Our focus was tactical not cultural. And the Commandant wanted "Values imprinted on their souls."

We were debating whether to cut recruit training to eleven weeks. And we had to decide whether the "bill payer" was the SDI [senior drill instruction] and drill time—those intangibles. But those are the times when values based training is taught. The guided instructions are all about that the DI can see the young recruit, "Does he get it?" So the Commandant kept twelve weeks and said we're actually going to cut some other things to add more time for VBT. . . . We created a values evaluation card for the Crucible. Technically you can fail the Crucible on values.

While the instructors at the recruit depots seemed confident that the current values based training program provides the basic physical and moral skills Marines needed to face combat, several of them expressed a concern that once Marines left the protected environment of the depot, this moral transformation was difficult to sustain. As Lieutenant Colonel (ret) Custer observed, "Our biggest challenge is sustaining values and character development. We've got a society that is morally at an interesting place—I'll leave it at that. The challenge is once your recruit leaves the protected environment, they go back into the large environment of the United States. And even the Marine Corps environment is struggling with these issues."

In 2012, responding to a new round of embarrassing incidents by Marines and other U.S. services in Afghanistan, General Amos, the 35th Commandant of the Marine Corps, mandated a new ethics training curriculum designed and led by the LeJeune Leadership Institute in Quantico, Virginia. The eight-hour class consists of lectures on the social, cultural, and psychological factors that lead to moral lapses in decision making; videos of such famous experiments as the Milgram study[14] and the Stanford prison experiment;[15] and a seminar discussion and analysis of the war crimes at Mi Lai, Vietnam. This course is mandatory training for all Marines, from recruit to colonel. Whether this course will become an annual required training event for all Marines or "just another shot in the arm," as one Marine commented to me, remains to be seen.

As I have discussed in this chapter, in the early years in Iraq Marines responded and adapted rapidly to the immediate social and cultural challenges they faced. Their ideals of "semper gumby," their bias for action, and their focus on mission accomplishment all were positive cultural factors that enabled many Marines to quickly grasp the importance of culture and language skills in theater. However, their rapid and simplified solutions to the situation also led to a confusing array of language and culture programs and an inconsistent vetting and training of instructors and translators. Equally if not more problematic, early aggressive and damaging actions toward the civilian population resulted in a backlash of hostility from Iraqis and a prolonged conflict. By reinforcing cultural ideals of honor and history through a comprehensive values based training and ethics program, Corps leadership has sought to remedy some of these challenges.

However, because Marines saw the initial cultural problems as requiring a simple quick fix to address the current conflict, most solutions required no fundamental change in the Marine Corps cultural ideals or ways of conducting war. Unfortunately for the Corps (and the other U.S. services), as I discuss in the following chapters, the need for cultural skills in combat did not disappear after Iraq.

7 "The 80 Percent Solution"

(Mis)-Translations from SME to Marine

IN 2009, A POWERPOINT SLIDE, AFFECTIONATELY NICK-named the "noodle" or "spaghetti" slide, was circulated around Washington, D.C., and the Department of Defense. The slide, which attempted to summarize on one page—through a complex spaghetti web of lines and nodes—-the complexity of the social and cultural factors affecting the war in Afghanistan, brought quite a few smiles and comments, including a column in the *New York Times*.[1] Headed by the tongue-in-cheek title "We Have Met the Enemy and He Is PowerPoint,"[2] the column poked fun at the absurd lengths to which PowerPoint presentations had seemingly devolved in the military. "When we understand that slide, we'll have won the war," General McChrystal was quoted as saying upon seeing the slide—a comment that apparently brought laughter from his staff.

Curiously, while members of the American civilian and academic population laughed at the slide as a parody of how the military viewed the cultural issues involved in the war in Afghanistan, some of the Marines I was working with viewed the slide quite differently. "You know that noodle diagram that was published in the *New York Times*?" asked Major Neal, an intelligence officer conducting research on the situation in Afghanistan. "Well everyone on the outside thought it was ridiculous. But those of us on the receiving end actually found it useful in thinking about the issues at work in the situation." As part of my research, I found the two differing perspectives provocative—indicating perhaps two separate worldviews and understandings of both the concept of culture and its representation and analysis.

Equally significant, while this slide may have seemed extremely reductionist to outside scholars, from my research perspective it represented an enormous shift from the early years of combat in Iraq, when the cultural challenges had primarily been viewed as a rather simple language and ethics training problem. It was true that the slide did indeed still attempt to simplify the cultural issues in Afghanistan to a one-page PowerPoint. However, a quick glance at the immensely confusing lines and networks made it clear that, at this point, the U.S. military saw the problem as complex, interrelated, and requiring coordination among both military and nonmilitary organizations. It also suggested that the military had made a genuine shift in thinking about the problem: viewing the key to success in Afghanistan and Iraq as a *cultural and social* rather than a *linguistic* issue, and recognizing that these cultural factors needed to be addressed at the higher strategic and operational levels—as, for example, in General McChrystal's planning room—as well as at the lower tactical level of the day-to-day interactions of Marines on the ground.

While it is not quite clear at what point this shift exactly happened, certainly by the time of the "Anbar Awakening" in Iraq (2007–2008) Marines had grasped the importance of understanding the local tribal, ethnic, and sectarian patterns and dynamics of the conflict. Interestingly, this time period also corresponds to a sudden rise in the number of subject matter experts (called SMEs) who began working with the military.

As I explain in this chapter, the employment of academic SMEs signified a second approach to responding to the external cultural imperatives of the operational environment and policy pressures from Congress: *translation*. Realizing that cultural understanding required time, study, and experience, the Corps (which had none of these) turned to outside experts to quickly translate and explain the situation in terms that would make sense to Marines. This approach, like the earlier efforts to simplify the issue (described in chapter 6), sought solutions *outside* of the military structure, requiring few permanent shifts in the way the Corps conducted its operations.

Recognizing the need to work with other nonmilitary organizations and groups, SMEs were brought into the planning rooms in order to improve integration and understanding between civilian and military partners. SMEs were also brought on as advisors to commanders in theater, whether as native cultural advisors (CULADs) or members of the Human Terrain Teams

(HTTs). And finally, viewing virtual training as a cheap and easy substitute for large classrooms taught by inconsistently skilled native-language speakers, SMEs were hired to develop language and culture modeling and simulation programs.

Ironically, although most SMEs and Marines technically came from the same cultural background (the United States), the reality was that each viewed the cultural aspects of the conflict in significantly different ways. Frequently, Marines assumed that SMEs working with them shared the same values and goals of accomplishing the mission quickly and effectively, asking them to provide information and analyses in ways that watered down the value of SMEs' input. Not necessarily understanding the language and culture of the Marine Corps, however, SMEs' assistance—whether computer programs, research, or analysis—also often failed to explain the cultural issues in ways the Marines expected, providing limited value or applicability for the Marines. Culture, then, frequently became "lost in translation" despite the many millions of dollars invested in SME-led programs and projects.

Placing Culture in the Planning Room: Green Cells and SMEs in the Joint Space

The windowless room in the Marine Corps Battle Staff Training Facility (BSTF) looked to me just like something straight out of a *Star Wars* movie. Five enormous video screens filled the walls. A colorful map of the fictive country of Indolaysia, showing its mountains, rivers, roads, rail lines, and ports, filled the two corner screens. Three other screens portrayed a second map showing the same landscape but overlaid with strange symbols—colored rectangles with X's and circles inside, black lines connecting the rectangles, and big blue arrows emanating from the boxes and converging on a set of red arrows. Pasted on the walls between the screens were sheets of butcher block paper and computer printouts with diagrams and charts providing a dizzying array of information: anything from the location of police stations, names of local mayors, tables of local imports and exports, plans of sewer systems, and diagrams of relationships between known insurgent groups in the area.

Below the screens and charts, a large, U-shaped table lined with laptop computers and stacks of papers followed the walls of the room. Along the table, twenty-four Marine majors and lieutenant colonels sat hurriedly working on their computers in front of them. Every now and then one of them would stand up to hand a paper or diagram to his colleague and speak in rapid but quiet tones. At the back of the room stood several colonels, chatting unconcernedly to a couple of civilians, who stuck out distinctly in their suits and ties. Seated at a separate table in the middle of the "U," two general officers shuffled a stack of papers, shared a quiet joke, and stared periodically at the five enormous screens on the walls, waiting.

The screens flickered. In front of the screens stood a Marine lieutenant colonel, slightly tense in the shoulders but composed.

"The slides?" he queried a Marine sitting at the table to his right.

"IPB Indolaysia" flashed on the three primary screens.

The lieutenant colonel glanced at the two general officers in front of him. They nodded.

"Generals. Colonels. This is Lieutenant Colonel Hastings. My team and I will be briefing you on the cultural IPB [intelligence preparation of the battlefield] for exercise Pacific Challenge Phase IV. There are twenty-eight slides, and approximately forty additional back-up slides. The brief should take approximately two hours."

Prior to my arrival at the Marine Corps, I somehow imagined wars to be fought by a large cadre of brave weapon-brandishing young men led by a brilliant gray-haired general smoking a cigar, who barked out orders to his courageous troops on the slopes below as he overlooked the unfolding battle. However, the reality for most U.S. military officers is far less glamorous. Contrary to the Marines' self-identity described in the first part of this book, rather than storming the slopes of Iwo Jima, most Marine officers spend a large percentage of their career shuffling papers, creating PowerPoint slides, and presenting their estimates in briefs (reports) to generals in windowless rooms. A news spoof about General Mattis's supposed retirement in the *Duffel Blog*[3] (the military equivalent to the *Huffington Post*) reveals this tension between Marine officers' self-identity as "hard charging devil dogs" and the reality of endless days spent in briefings:

> General Mattis, 62, is circumspect about his upcoming retirement. "I postponed my retirement to lead CENTCOM, but it's just time to move on. Years ago I was called to serve my country by means of conferences and briefings, but I feel like, after forty years of that, I've done my duty. . . . When you're young, it's nothing to sit through even a four hour meeting. You think "oh, I'm only thirty, I've got plenty of time left to spend killing the enemy," or "it's okay, this asshole wasting

my time will get s—-canned if I just outlast him, and then I can get back into the field." But the hours turn into days, the days turn into weeks, then one day you look up and you're sixty, still sitting in these goddamn meetings.

While most Marines (like the fictional General Mattis interview reveals) probably chafe in frustration at the long, often deadening hours spent in planning and briefing for military operations, today's modern warfare is far too complex for general officers to be able to simply stand on hills and oversee the battle from above. Contemporary military operations require the support of a large, skilled staff that researches and updates the general on all aspects of the problem. Whether conducting a combined air-ground-sea attack on Iraq during Operation Iraqi Freedom (OIF) I or providing stability to the same country through combating local insurgents, reconstructing roads and schools, and training police and military officers, Marine operations today demand immense research, planning, and coordination with civilian and other non-military groups and individuals.

Reflecting this multidisciplinary approach to resolving conflict, today's military officers at Marine Corps University not only study military history but also political science, international relations, ethics, law, and even anthropology to understand the complex factors they might face in their operations. Equally important, they also receive practical experience in military planning through simulated wargames such as the fictive "Indolaysia" exercise described above. Part of the purpose of these wargaming exercises is to teach Marines and their service partners how to communicate and work together effectively on large staffs consisting of a huge variety of military and civilian members—a difficult task given the many different perspectives and ways of conducting business posed by such joint staffs.

One of the more significant steps that the Marine Corps has taken to incorporate cultural factors, joint partners, and SME input into the planning process has been the recent development of the Green Cell. Released as doctrine in 2011, "The Green Cell is a commander's planning tool to help better understand the environment he is in with respect to the indigenous population, to better understand the nature of the problem he is facing writ large, and ultimately to make more informed decisions."[4] In conventional Marine Corps planning, combat operations have been viewed as consisting of two opposing groups: the friendly forces (blue) versus the hostile forces (red). The Green Cell[5] adds another category of actors to the equation: one that can be neutral, friendly, or hostile alternatively. This category is the population of the country. Although the Green Cell as a concept does reflect a potentially significant shift in Marine ways of approaching conflict, its implementation is another story, for like all the other planning groups on the team, the Green Cell must produce its analyses in the accepted language of the Corps: PowerPoint.

PowerPoint, Maps, and "Human Terrain"

Today, whether Marines or other staff members like it or not (and many do not), the primary medium for communicating in the military is by providing summaries of one's work through PowerPoint. As General Taylor advised the students in the Indolaysia exercise, "[You must] become a student of briefing. Man has evolved to PowerPoint. That's the way of military life. We use briefing."

From a cultural perspective, the use of PowerPoint to communicate in the military reflects a uniquely visual way of thinking about and conceptualizing the world. While PowerPoint is the dominant method used to impart information and understanding, it is only one of many visual ways that Marines and other members of the military analyze and evaluate issues. In my work, both at the university and with Marine Corps organizations, it was expected that we would not only use PowerPoint presentations but also maps, diagrams, charts, tables, graphs, videos, and even dynamic fading and appearing pictures or words to represent the problems we were discussing. In helping me to prepare my lectures during my first weeks at the university, for example, Dr. Green explained to me, "Visual representation of information is essential—[for example, in this lecture I use a] hierarchy map, networks maps, photos, videos, and aerial photos."

In fact, any method that communicates visually is considered effective. In remote deployments, often flip charts and butcher block paper are the main forms of communication, rather than the fancy advanced technology described earlier in the BSTF. On a university field trip to Gettysburg, for example, electronic representation of information was not possible. Instead, students carried an easel across the fields, setting it up to display maps and other diagrams about the famous Civil War battle movements at each historic location. The site, the battles, and the historic events that occurred there were all interpreted and analyzed through these maps and diagrams. In commenting on the presentations, one of the students stated to me, "We're very visual. Notice no one moves away from the map."

This emphasis on conceptualizing the world visually was explained to me by Colonel Simons: "We just love to look at things in maps and pictures. That's the way we were brought up. We just love that." Elaborating further, he explained how the physical environment—or "terrain"—is a central part of the way that Marines look at the world. "We think about everything in terms of terrain. I can't drive home without thinking of everything in terms of terrain—how the road rises, where the river is. It's something that Marines can relate to."

In traditional battlefield analyses, the military preference for information that is transmitted visually is quite logical. Conventional military battles are fought on clearly defined physical spaces (or terrain), which are well suited to analysis using geographic techniques such as mapping. Not surprisingly, then, early on in the culture venture (and congruent with this visual way of thinking about conflict), culture became translated into the military metaphor of "*human terrain*." Like physical terrain, which could be mapped and understood graphically, putting humans *on the map* made sense.

Explaining why "human or cultural terrain" was a useful way for the military to think about people who were in the battlespace, Major Neal stated, "Culture is a tool that can be used in any environment. That's why we focus on cultural terrain as a concept that fits within the military. [It provides a] process of how to use cultural information in a diagram we are familiar with." In a demonstration of the "Mapping the Human Terrain Project," called Map H-T, one of the leaders of the project emphasized its value to the military by stating, "[Our] product is a battlefield information sharing system that shows cultural information on a map. It helps inform my planning on a macro level." He continued by noting that this system would help "to get your arms around all that plethora—to get human data into a format that is useful. How do you help manage what the commander on the ground needs to know? [With this product] you can visually predict in layers." His comments thus emphasized the way that mapping "human terrain" could help commanders quickly grasp and constrain the complex, "soft," and unfamiliar concept of culture in a structure that was familiar to them.

Although "human terrain" is a metaphor that makes sense to a military culture that views the world through maps, this two-dimensional, static way of conceptualizing culture produced some rather unusual interpretations of human behavior (from my academic and scholarly perspective), especially when used to analyze a nonconventional battlespace. This strange "mis"-translation became very clear to me during a final cumulative wargaming exercise at Command and Staff College (CSC) called "Nine Innings."

Deviled Eggs?: Eggheads and Devil Dogs Share Plans

Requiring the preparation of a fictive operational/strategic plan for stabilizing the Philippines, the final wargame of Nine Innings at CSC involved the entire school of 200 students plus faculty and numerous outside SMEs from Marine Corps University. Over a two-week period, students were split up into planning cells (or groups) and were assigned different issues to research and report on to the commanding general for the exercise. Cells and teams were distinguished by colors (blue for friendly forces, red for hostile forces, and green for the "neutral" population). Indeed, the military convention of labeling various actors in the battlespace by a color rather than a number or a name immediately emphasized the visual nature of the exercise.

Most interesting from a cultural point of view, however, were the products of the exercise. Each day, students produced PowerPoint slides that visually summarized their findings. Economic, political, social, cultural, ethnic, linguistic, and militarily strategic information was depicted through colorful graphs, charts, and maps. Words were primarily reduced to a few key terms below the slide, definitions of symbols, or, occasionally, bulleted lists.

Students creatively found ways to summarize the information they had collected over the day, using signs such as arrows, lines, color coding, circles, flags, and numerous other symbols that were initially unintelligible to me. Many of these strange symbols, it turned out, were a standard set of military symbols that all students study in the joint Marine-Army publication *Operational Terms and Graphics*, best known as FM101–5–1.[6] Like the symbols one learns for reading geographic or nautical maps, these symbols are part of the military language and integrally tied to mapping functions.

A particularly interesting case example was the PowerPoint produced by the Green Cell with which I was working. Students in the cell were assigned the task of briefing the commanding general on the ethnic and linguistic composition of the Philippines and then relating this composition to levels of income or poverty. The students in the cell were initially rather overwhelmed by the ethnic and linguistic data: the Philippines consist of well over 100 ethnic and linguistic groups scattered over more than 7,000 islands. Furthermore, maps of poverty and income levels were incomplete across the country, making any kind of meaningful correlation unlikely.

To my great surprise, on the day of the briefing the students had succeeded in reducing an immense amount of complex data into one simple slide. The slide agglomerated the more than 100 ethnic and linguistic groups into three simple categories. Then, on the basis of the simpler, reshaped ethnographic landscape, a map of poverty across the Philippines was overlaid upon the ethnic data, suggesting a weak relationship between the two. From my perspective as a SME, it seemed I had failed in assisting the students in conducting an effective analysis of a complex situation: one where the culture and characteristics of the local population were not fixed and easily measurable and where meaningful relationships between factors would require careful statistical, not visual, analysis. However, the students' PowerPoint slide was received quite well by the military leadership for the exercise, who saw the slide as providing exactly what they needed.

What had happened in the translation? Why would visual analyses such as the noodle slide described at the beginning of the chapter or the Philippines cultural summary in Nine Innings "hit the target" for its military viewers while seeming extremely simplistic and reductionist to me or other academics? As I learned through my later observations and interviews, from my students' point of view, it would have been professional suicide to provide the general officers with reams of confusing and even conflicting information (which is what I, as a scholar, thought appropriate). Due to Marine Corps (and general military) cultural ideals of decisiveness and the ability to respond quickly in ambiguous situations, my students understood that successful communication and briefs to the general required them to be succinct and provide only the most essential information in a short period of time.

"Give It to Me Barney Style": Time, Speed, and Simplification

In his practical, terse, and direct way, General Taylor explained to me one day why the military did not have time to spend hours researching and analyzing issues. "Part of the problem we had at [a joint headquarters command] was that we just sat around admiring the problem. But we who are in the military have to do something. We have to go out and deal with the tsunami, the insurgent, whatever. We can't just sit there."

Given the rapid "churn," or tempo, of operations when deployed, Marines often have very little time to conduct anything more than a quick surface assessment of an issue. Describing the frantic pace and time pressure on one of the organizations evaluating cultural issues in Afghanistan while he was deployed, Lieutenant Colonel Lyons observed, "Their problem was partly that they were expected to put out a product each day. I don't blame them, if I had to put out a report every day that's what I'd do [simplify the issue]. So you end up with a report that's an inch deep and a mile wide. It's a check-in-the-box."

Echoing this concern with lack of time, at one of the warfighting exercises a Marine intelligence analyst commented to me, "We don't have time for the complex analysis. So what we need to do is extract the important information, download it to a system, and then input it into a map. We're all about maps."

This sense of urgency and the need to take action and be decisive often leads to a disconnect between Marines and the Ph.D.s working with them, who are often accused of losing sight of the main objective. In leading a working group composed of Marines and academics, Colonel Irons reflected this concern: "I don't want this to be an academic discussion—no offense to you academics—but those tend to just get going and going and [lifts arms wide]. We need to get something done. So if we can try to focus the discussion and get some outputs."

Conversely, the expectation that SMEs could provide a hastily assembled, simplistic analysis of a problem also led to frustration on the academics' part. As Mary Fielding commented to me about her work on a Human Terrain Team in Iraq, "PowerPoint briefings. You really have to know the language of your audience. Sometimes you would be asked to provide one slide with five points and only two seconds to tell what it means. It was an incredible experience to me to walk in and have someone from the Marines tell me that 'You have ten minutes to tell me how you are value-added, and if you don't you can leave.'"

Marines and academics not only have different expectations regarding the appropriate amount of time necessary to research a problem, but, as Fielding notes above, they both also diverge on the kinds and amount of information that needs to be provided. Most of the Marines I interviewed subscribed to the KISS (Keep It Simple Stupid) approach to communicating information, especially in briefs to general officers. In numerous presentations, I have heard the commanding officer chide a presenter who was talking too long, or providing too much information, state, "Give it to me Barney style"—indicating he wanted a very basic simple explanation. "I think less is more," stated one of the lieutenant colonels regarding his brief on culture for one of the Pacific Challenge planning exercises, "if I put too much in the slide I'll confuse them."

Frequently, because so many people are contributing an immense amount of information in a brief to a general officer, even if they would like to include more information, both Marines and SMEs are rarely given the opportunity to provide more that one PowerPoint slide on their work. During the CSC Nine Innings exercise, for example, students in the Green Cell were expected to summarize an entire day's worth of information and research into one slide for their brief to the commanding general. As one student explained, "We cannot afford to get down into the details. We have a very defined endstate."

Limited time and slide space, however, are not the only reasons that Marines might condense the information into a few PowerPoint slides. As Major Davis observed about his role on a planning team in Iraq, "If you can't put it up on a wall and you can't visualize it—it goes away. I had a forty-page document but if I hadn't put it up on the wall, they wouldn't have understood any of it." The need to present simple, concise, visually clear information was also expressed by one of the majors in a Nine Innings exercise: "How you display the system is very influential in understanding it. We try to map the problem or create a framework to constrain the problem."

In another conversation about a culture training brief, Lieutenant Colonel Jones stated, "A picture is worth a thousand words. That's ten times more important for the operator. Yes there's more details [that should be included] and a lot of information is left out. But the operator isn't going to read that."

However, while virtually all of the Marines I talked to were clear that the military system left little time or space for complex analyses, several of them expressed their concerns that such analyses were not always sufficient. In the Nine Innings exercise, for example, one of the team leaders expressed his difficulty with reducing a large amount of cultural information into a slide: "Graphic tools are great, but we're having the problem of trying to depict it. We can conceptualize the problem, but we're having difficulty representing it." Similarly, a Marine major who was working on a computer modeling program for tribal structures in Iraq also observed, "Demographics are changing rapidly. Tribal movements, people movements change everything in six months. It changes daily even. To try to map it is improbable."

And in discussing his experiences in the brigade command staff in Afghanistan, Colonel Chase remarked about the cultural aspects of the situation, "I think we're too quick to propose solutions before we understand the problem. I think we need to be taking more time, studying and understanding the problems and making less PowerPoints."

Planning teams are not the only place where Marines and SMEs faced challenges in translating between two different worlds—between the perspectives of those with "hard skills" (tangible, concrete visible skills such as shooting a gun) and "soft skills" (those intangible, hard-to-quantify understandings of human relationships and interactions). This difficult dance of learning how to communicate and work with each other has also been particularly challenging in theater, where Marines and other service members must work together with government and private civilians, academic SMEs, and the host nation's government and security forces to understand and resolve the human as well as military issues in the conflict area or battlespace.

Culture SMEs in Theater: The Ethical
Challenges of Human Terrain Teams

In addition to hiring interpreters to assist with language translation in Iraq and Afghanistan, the U.S. government and military engaged a second set of academic experts to aid in understanding the social and cultural issues in theater: SMEs. While these SMEs have worked in many roles, ranging from positions on the joint staffs to contractors in the field, the most publicized and controversial engagement of SMEs has been with the Army's Human Terrain Teams.

Started in 2007, the Army's HTTs were intended to provide a new critical cultural capacity to the military battalions stationed in Iraq and Afghanistan. According to the Human Terrain Team handbook, "HTTs are five to nine person teams deployed by the Human Terrain System to support field commanders by filling their cultural knowledge gap in the current operating environment and providing cultural interpretations of events occurring within their area of operations."[7] These teams were expected to go out among the population and assess the situation from the local people's perspective, reporting back to the commander on what they had seen and heard.

The Marine Corps never actually funded or deployed any of its own Human Terrain Teams, which were under the direction of the Army. However, a number of Marine units, working in conjunction with the Army, were assigned to work with HTTs. The HTTs received considerable negative attention in the media from some members of the American Anthropological Association (AAA), who complained that the use of anthropologists on the teams violated ethical standards of the discipline.[8] (An official report on the issue by the AAA found that these teams were not conducting anthropology, thus sidestepping the conflict.)[9] Ironically, given that most of the complaints came from members of the AAA, very few cultural anthropologists actually ever deployed on the five-to-nine-person teams, which were typically composed of a team leader, one-to-two social scientists (who typically held a bachelor's, master's, or doctoral degree in *any* social science discipline, not just anthropology), two-to-four analysts, and a research manager. The majority of the team members were retired or active duty military, with an occasional political scientist or international relations specialist as the "social scientist" rather than an anthropologist.[10] These teams were intended to assist military commanders in understanding the cultural situation in the area.

One of the major challenges for these teams was that the social scientists were ostensibly hired to provide a general understanding of what was going on in the area—offering openly accessible information (not intelligence) about overall attitudes and concerns faced by the population. However, to the Marines and soldiers working with these teams, the academics on HTTs were typically seen as "combat assets," assisting the commander in applying cultural knowledge to the battlespace. Thus while the social scientists on these teams tended to view their role as providing research data and studies of the local communities in order to advance the military's basic cultural understanding and knowledge, Marines and soldiers working with the social scientists tended to see them as just another kind of "intelligence asset"— providing secret information that could be used to locate and target possible insurgents and prevent their hostile activities.

This differing interpretation of the use of cultural information from SMEs was described by Colonel James, who worked closely with an HTT during his deployment to Iraq. "[Sometimes] there were translation issues. One afternoon I was trying to work out a predictive tool, so I needed info. I looked over the room and asked Mary [the social scientist on the team], 'Can you give me this information?' She said, 'I can't do that. That would be targeting.'"

The translation problem worked both ways, however, as Mary Fielding (the HTT social scientist mentioned above) later added, "[I have to emphasize the] importance of understanding the culture of the military you're working with. There may be someone who knows more about the subject than you, but if you are able to talk to the commander and explain in a way that he wanted, you would be the one listened to. There were a lot of social scientists with Ph.D.s who were walking around telling everyone about their degrees and their universities and how important they were. And they would talk to the military commander as if he was barely out of kindergarten. It turned out that three of the commanders had Ph.D.s from Ivy League universities." She continued, noting that over time SMEs and the military officers slowly began to understand what the other was trying to accomplish. "What was important," she stated, "was when they started trying to understand academics. We went from eggheads to becoming a member of the team."

Virtual SMEs: The Failure of Language and Culture Programming for the Corps

While SMEs and the Marines in theater and on the planning teams were busy churning out PowerPoints and daily briefs in response to rapidly changing daily crises, back in the United States, the Department of Defense and the U.S. military services did attempt to address the larger issue of undertaking sustained research on social science issues related to the current conflicts. Recognizing a need for more in-depth research and understanding of the social and cultural challenges facing the military, many of the services developed specific lines of research funding for social science projects and established an immense variety of new culture programs and research centers.[11]

Given the Marine Corps' relatively small budget, its "every Marine a rifleman," its "can-do" approach, and its bias against specialization, only a handful of the academic SMEs hired by the military were actually funded by the Marine Corps. "If we don't get SMEs, fine. We'll do without them," stated one member of a culture simulation program for the Marine Corps. As General Taylor noted, "There is a SME out there who is the expert on almost anything. But there is a tendency in the Marine Corps to think that if we work harder and roll up our sleeves we'll get the job done [without them]. That might be so, but it would be better to go out and get someone who could help." As a result, much of the research and development funding for these research SMEs has come from the U.S. Navy, Army, Air Force, and DoD rather than the Marine Corps.

Initially, a number of Marines expressed enthusiasm for the more practical and applied forms of culturally oriented research and especially for the development of computer programs and software. Because technology has solved so many of the military's challenges in the past, there was a belief by some (including social scientists who benefited from the funding) that computers and technology could be used to solve anything from nuclear war to irregular warfare. As one research scientist stated rather optimistically in a presentation to a group of Marine leaders regarding a new program, "If you can do nuclear science on a computer system, you can put human science on a computer system."

Since the new generation of incoming Marines has grown up in a world of video games, movies, and iPods, it was assumed that computer games and programs would especially appeal to this new audience. For example, a contractor promoted his work on a new video software program for culture training with the argument that "the young lieutenant will play with the game, whereas he will nod off in the workshop." Captain Nash, who was preparing his company for deployment to Iraq, stated of a tactical language program the Marines were using (with avatars moving and speaking on the screen), "You put a Marine by himself and say, 'Hey do this tactical language program,' they're all gonna learn something. . . . Marines are very visual, doing type learners. So anything that goes in that realm, where there's still things like the mazes—where you have to turn left, turn right to get the guidon [unit flag]. I mean that stuff works well with the Marines."

Since the Marine Corps was struggling with the challenge of how to offer classes that would fit into Marines' schedules, an obvious solution was to create computer disks (CDs) with language and culture classes and video games that Marines could then take home or to a computer on base. Reflecting the hope that computer language and culture learning software and games would provide an easy solution, millions of dollars were poured into funding the development of this software by many of the services, including the Marine Corps. Investing considerable funds on this premise, the Marine Corps purchased a number of online language courses and set up top-of-the-line, high-tech computerized buildings on every Marine Corps base. These buildings, called language learning research centers (LLRCs), were intended to provide a space where Marines could come in and practice their language and culture skills on the computer, as well as offer instructional space for predeployment training.

Within three years of their installation, however, it was clear from the usage numbers that few Marines came to the buildings to practice their language and culture skills on the computer. A similarly low number of users were logging in from home or on base to access the online language software programs that the Marine Corps had purchased. In my interviews and the

Center for Advanced Operational Culture Learning survey, several explanations arose for the low interest in the computer programs and LLRCs.

First, as several Marines I talked to suggested, they had limited free time, and spending it on language and culture programs was not a priority for most. "We get all this stuff on CDs. But so, OK—Marines are not going to do this on their off time most likely unless we force them to do it," stated Captain Rhodes of Battalion 1/7. "There's limited assets that we can say all right—ten of you sit on the computer, the other twenty of you go clean weapons." Likewise, Major (ret) Jefferson observed, "First-person shooter games. You go home, whack someone, shoot the bad guy. Your sergeant will do that. But going home to play a game where you shake some hands, drink some tea . . . [shakes his head]."

Second, and perhaps more important, many of the Marines in the study believed that to understand culture or learn a language, you needed to speak and interact with a real human being, not a computer. For example, Captain Nash, who explained above how the tactical language training program could have value for his company, continued by adding a caution to his comments. "If it's the civilian guy who just runs the program and can turn the program on and log on and off, you don't get as much value as if you had, say an actual interpreter. So . . . it's kind of sinking things. Like we've got how many Arabic-speaking Iraqis over across the street from Mojave Viper? You know, [we need more interpreters] having more interaction with the battalion outside of Mojave Viper—along those lines."

This notion that culture and language skills were best learned by interacting with real people from the country was also mentioned by respondents to the CAOCL survey. One Marine corporal and infantry rifleman stated, "As with a lot of languages, the particular region of Iraq that we went to, there were significant variations between what we were taught and what the correct way was. . . . Once there, we had Iraqi police with us. Through them, I learned plenty to have conversations with them in their language. That greatly improved the effectiveness of the operation. I learned more by talking to Iraqi Army soldiers than most Marines learned and retained through our Rosetta Stone language courses that a few were sent to."

"The Marines are getting force fed through MarineNet [the online computer training system] and not through enough interaction or live training," stated a Marine gunnery sergeant and infantry unit leader on the survey. "There definitely needs to be more foreign language/cultural instructors

TABLE 7.1 Preferred culture training methods

Preferred method	N	Percent
1. Small group discussion with a SME	609	25.4%
2. Lecture from an instructor	556	23.2%
3. Immersion experience (e.g., Mojave Viper)	503	21.0%
4. Scenario-based seminar (case study, vignettes)	262	10.9%
5. Computer-based training	154	6.4%
6. Reading materials (smart cards, etc.)	135	5.6%
7. Video instruction	75	3.1%
8. Gaming materials	66	2.8%
9. Virtual learning environments (emails, chat)	36	1.5%
Total	2,396	100%

SOURCE: CAOCL survey

to meet all of the requirements. . . . Having Marines sit around a computer screen is not the proper way of teaching language and culture." This view that culture and language are best taught by doing—by interaction and role play with heritage speakers from the country—was also supported by the statistical data from the CAOCL survey.

As Table 7.1 indicates, the top three learning methods for culture training all included interaction with people from the culture. Only 6 percent of the 2,406 Marines in the survey preferred computer-based training, and only 3 percent selected video games as their primary learning style. Computer-based training was rated more highly for language learning, with 20 percent of the survey respondents selecting this method (see Table 7.2). However, live instruction with a native speaker or through immersion was still preferred by almost 50 percent of the respondents.

Despite the immense amount of money and effort dedicated by the Marine Corps and numerous SMEs to developing culture and language software, the project has more or less been abandoned. In 2012, six years after LLRCs were built, more than a quarter of the centers were shut down, with the remainder slated to be phased out and/or sold to the bases for other uses in the next two years. All contracts for online language software were also canceled. As one of the leaders at CAOCL explained, "It's really the story of language-learning software. It's not an effective tool for Marines to use."

Given the Marine Corps' cultural emphasis on learning by doing and Marines' view that the best information comes from "boots on the ground," the failure of computerized language and culture programs is not surprising.

TABLE 7.2 Preferred language training methods

Preferred method	N	Percent
1. Language instruction with a live instructor	660	27.4%
2. Web or computer-based instruction	502	20.9%
3. Immersion experience (e.g., Mojave Viper)	461	19.2%
4. Private language instruction	230	9.6%
5. Scenario and role play	157	6.5%
6. Portable technology (iPod, iPhone)	132	5.5%
7. Gaming materials	111	4.5%
8. Printed phrase lists	79	3.3%
9. Language labs	74	3.1%
Total	2,406	100%

SOURCE: CAOCL survey

Furthermore, the lack of time and Marines' view of themselves as strong physical warriors (rather than diplomats) make it unlikely that most would voluntarily study language or culture in their free time.

The Marine Corps' unsuccessful experimentation with computerized culture and language programs offers some hard lessons about funding programs based on stereotypical beliefs about "young Marines from the computer generation." Without examining the routines, values, and concerns of Marines to understand their world and culture, millions of dollars were unnecessarily expended on a target audience that did not want and/or was unable to use the program as delivered. This, of course, is a lesson that has been learned over and over again in many failed cross-cultural projects, from international development programs to efforts to improve education and health care in impoverished U.S. communities.

The difficulties faced by Marines and SMEs in understanding what the other needed in order to translate successfully between the different cultures provide a fascinating challenge. As the examples above illustrate, while Marines and SMEs may both originate from the same American culture, that does not mean they view or interpret the world in the same way. Paradoxically, the problem suggests that cultural challenges may not simply exist "out there" between the United States and other countries, but between our own subcultures: whether they are service cultures, NGO cultures, government cultures, or civilian and academic cultures.

8 "Where's the 'So What'?"

Processing Culture at the Center for Advanced Operational Culture Learning (CAOCL)

The three double-wide, gray metal-sided trailers sit less than a hundred yards away from the train tracks on the Marine Corps Base in Quantico, Virginia. Tucked behind the back of the U.S. post office, facing a parking lot of postal trucks on one side and a chicken wire fence separating it from the Amtrak and Virginia Railway Express (VRE) train stop on the other, the Center for Advanced Operational Culture Learning (CAOCL) is virtually impossible to find without a GPS. In the humid and hot Virginia summer, the windowless conference rooms in the trailer become stifling, encouraging short meetings with all who dare venture to find the Marine Corps culture center. In the winter, employees must walk gingerly on the metal walkways between the three colocated trailers, lest they slip on the snow and ice. Despite my regular visits to the trailers, after six years I still cannot shake the sense that somehow the odd complex reminds me of the rapidly constructed immigrant settlements circling the outskirts of Tunis—temporary illegal and unwanted settlers tucked away from view.

Entering the trailers, I am confronted with the scent of saffron and lamb, making me feel all the more convinced that I am actually in some faraway Middle Eastern country. The Afghan culture and language instructors are heating up their noonday meals, and generous as always, have laid out a mouth-watering spread on the lunch room table.

"Eat, come eat with us," Muhammad pats an empty seat next to him.

I look out at the sea of international faces seated around the table, each smiling and indicating the dishes they have contributed to the table: Emira, CAOCL's Arabic instructor, has brought hummus; Sumayia, the East African curriculum writer from Ethiopia, shows a spicy meat dish; Hassan has laid out a Moroccan couscous; José, CAOCL's Latin American officer, is eating a Columbian meat pastry; and Megawati, who is wearing a beautiful flowing purple dress, has brought her famous Indonesian ginger tea. I cannot help but smile at the contrast between my office at Marine Corps University, where I teach khaki-uniformed military officers who speak in quiet tones in the severe hallways, and the heady scented buildings and colorfully dressed employees at CAOCL.

FROM THE VERY FIRST DAYS OF CAOCL'S OFFICIAL founding in 2006, it has been an anomaly. An incongruent addition to traditional Marine Corps ways of conducting business, the Marine Corps' culture center has been a source of debate, needing to justify its value to the Corps ever since it was established. Yet what is perhaps most surprising, even miraculous, about CAOCL is not its predominant workforce of multicultural, multilingual civilians[1] with few Marines in sight, or its odd location and hodgepodge buildings, but that despite its temporary structure and unusual purpose, more than ten years after the initial invasion of Iraq, CAOCL still continues to have a role in the Marine Corps.

As I discuss in this chapter, CAOCL's continued existence can be attributed in large part to its success in *processing* the new, strange, "foreign," non-Marine cultural policies into accepted Marine Corps ways of doing things. Just as factories take raw materials (for example, meat) and subject them to a process that creates a new standardized product (such as neat, equally sized sausage links), so, too, has the Corps taken a complex, nonlinear, "touchy-feely" concept—culture—and subjected it to a standardizing process that produces something that "looks, tastes, and smells Marine." Rather than challenging existing methods and approaches to Marine Corps training, the "culture concept" has been reshaped and reworked to fit into Marine Corps organizational processes, identity, and ways of seeing the world.

The result is fascinating—a standardized "one size fits all" culture and language training framework based on "the Five Dimensions of Operational Culture"[2]–which can easily be adapted to any culture or country to which Marines may suddenly deploy. The program's simple, standardized format and structure can be taught to any Marine, regardless of rank, background, or skill level—compatible with Marine Corps ideals of "every Marine a rifleman." In the process, the complex, conceptual nature of cultural understanding becomes restructured into a mission-focused set of behaviors and actions that, with quick revision to fit the latest crisis destination, can be adjusted to any location to which Marines deploy. Culture becomes, then, "a tool in the kitbox"—a skill like shooting a rifle or flying an airplane—that Marines can use to achieve the mission wherever they may be sent. Reflecting this notion that culture is a manipulable object, Major Neal noted in an interview, "Culture is a tool that can be used in any environment."

Indeed, by restructuring culture and fitting it into familiar training processes, culture becomes conceptualized as analogous to other weapons that Marines use to achieve the mission. "The commander is using culture to shape his battlespace, just like he's using artillery, mortars," explained Lieutenant Colonel Jones, who was assisting with the CAOCL training program. "It's about mission effectiveness. Our mission is to close with and destroy the enemy. Not only is culture a direct fire weapon, it's an indirect fire weapon." As a result, culture and language become integrated into the normal processes of Marine preparation for deployment, requiring training in order to master this new skill—just as one masters shooting a weapon through training.

CAOCL's role and purpose thus have become defined in Marine Corps terms, using Marine Corps metaphors and ways of thinking about the world, ensuring its continuing survival (to date) despite its tumultuous history.

Survival of the Most Culturally Astute: The Story of CAOCL

As discussed in chapter 6, Marines initially responded to the cultural challenges of operating in Iraq with an enormous variety of creative solutions to the problem. The result was an incredible hodgepodge of predeployment training and in-country[3] programs ranging from culture classes provided by Iraqi expatriates, to Jordanian interpreters acting as translators and teachers in-country, to language classes for Marines at the local community college. Within a year of the invasion of Iraq, it became clear that this "do-it-yourself" approach to culture and language had created a very inconsistent patchwork of solutions and approaches. So in 2004 Training and Education Command (TECOM) assumed responsibility for all aspects of training, including culture and language predeployment training.[4] Rather than let each battalion commander or the various MEF commanders decide what training Marines should receive, TECOM mandated a standardized set of training requirements for all deploying units. As one of the staff members in TECOM at the time explained, "Before OEF II [Operation Iraqi Freedom II], predeployment training was a MEF responsibility, not a headquarters [responsibility]. During OEF II we started to see battalions coming through with different training. So in late 2004, headquarters TECOM took it over."

One of the more important results of this new approach was the creation of a formal culture and language center, CAOCL, established to oversee, provide, and ultimately standardize the training of culture and language for all Marines across the Corps. According to its official charter, the mission of CAOCL was to "serve as the central Marine Corps agency for operational culture training and operational language familiarization programs and issues within the Doctrine, Organization, Training, Materiel, Leadership and education, Personnel, Facilities (DOTMLPF) process in order to synchronize and provide for training requirements."[5] (It is important to note that at the same time that CAOCL was founded, the Marine Corps Intelligence Activity [MCIA] was also shifting its focus to providing cultural analysis and training for the Marine Corps from an intelligence perspective.)[6]

In 2005, culture and language training was still seen as an immediate solution to a temporary crisis: helping the Marine Corps deal with the hostile population in Iraq. CAOCL's temporary structure and personnel reflected this short-term effort at a solution. Initially headed by Colonel Jeffery Bearor, in its early formation phase CAOCL had only two other permanent government employees: a Marine lieutenant colonel and a civilian government employee who had expertise in Middle East studies. And the center had virtually no permanent government funding (its funding derived from supplementary, contingency operating funds dedicated to the conflicts in Iraq and, later, Afghanistan: termed "OCO funds"). As a result, initially CAOCL could only hire contractors—individuals employed by private contracting companies to provide short-term skilled labor to the government (the equivalent of temporary employees). Reflecting the rapidly created and temporary nature of the center, it was housed in a single aluminum-sided, windowless trailer in an unwanted and unoccupied space next to the train tracks.

Like all new start-up organizations, the first years of CAOCL saw immense growth. By the time I arrived at the center in the fall of 2006, CAOCL had approximately twenty-five employees. By 2013, its staff numbered over seventy people. Only five, however, were Marines in uniform; four were civilian government employees; and the remaining sixty or so were still short-term contractors. Visually emphasizing its continuing temporary nature, the center was still housed in trailers. But now instead of one double-wide trailer, three trailers stood behind the train tracks.

This expansion, however, masked a rather bumpy trajectory of fits and starts. In 2007, with the promotion of its director, Colonel Bearor (now retired) to the position of deputy director of TECOM, the future of CAOCL was quite unclear. For a period of time, the center was subsumed under another new start-up organization, the Center for Irregular Warfare (CIW). Then, in 2008, Major General Spiese (the new commanding general of TECOM) directed that CAOCL was to be separated from CIW once again. A new permanent director was assigned to CAOCL, Colonel George Dallas, who had just retired from the Marine Corps.

Within a few months of his arrival, Colonel (ret) Dallas realized that the existing confusing array of do-it-yourself culture and language programs developed during the early years in Iraq were harming the effort to integrate culture into Marine operations more than they were helping. Immediately, the new director focused his energies on creating a center that would provide training and programs that were relevant to Marines and fit into Marine Corps ways of doing things. As I discuss in the following sections, this was achieved by developing cultural training that would follow the standards and model of other training programs: "processing" culture to fit within accepted Marine Corps methods.

Although Colonel (ret) Dallas's efforts to "weave culture training into the very fabric of the Marine Corps" were largely effective, in 2011 the center faced yet another serious challenge to its existence. The newest commanding general ordered a study to restructure TECOM and remove redundant organizations. One of the organizations under scrutiny was CAOCL. Some organizations, such as the Security Cooperation Education Training Center (SCETC), disappeared. Yet after intense review, in 2012 CAOCL was given a permanent home under Marine Corps University. Placed under a securely funded and permanent institution, CAOCL appeared to have a much more stable future.

However, CAOCL's struggle for survival continues. The majority of the center's funding still derives from supplemental budgets (OCO, or operational contingency, funds) to support the conflict in Afghanistan. And as the Marine Corps and U.S. military continue to draw down from operations in Afghanistan, these funds will disappear. With the serious government cutbacks in 2013 due to sequestration,[7] there is little likelihood that CAOCL will receive new funds to replace those that were lost. Given that only four employees hold permanently funded government positions in the organization, it is likely that in the coming years, the remaining staff of approximately sixty temporary contract employees will be reduced dramatically.

Yet despite this gloomy picture, senior leaders have indicated to me that after the Marine Corps withdraws from Afghanistan, they believe that the center will have a continued (if diminished) role in the university and the Corps. Time, of course, will tell whether the Marine Corps' culture center was indeed just a short-term solution to an immediate operational problem or whether, in fact, the Marine Corps has truly adapted and accepted culture and language skills as a necessary part of contemporary training and education. To a large degree, whether CAOCL survives will depend on how well the organization can succeed in "processing" these skills to fit permanently within the Marine Corps culture. Three central ways that CAOCL has sought to achieve this is by creating programs that were operationally relevant; focusing on "learning by doing"; and standardizing the training so that it has been accessible to all Marines deploying anywhere in the world.

"We're Not Teaching Empathy Here": Operational Relevance

In the very early days of Marine Corps culture and language training, much of the emphasis was on basic customs, courtesies, and greetings—what Marines referred to as "do's and don'ts." These customs and courtesies briefs (as PowerPoint classes are referred to) might have filled the early concern of creating ethical Marines who would not do anything to stain the honor of the Corps (see chapters 4 and 6). However, by the time I had arrived in 2006, many of the Marines I spoke to expressed a frustration with this simplistic approach to the role of culture in military operations.

Describing this notion that such simplistic "hands and feet" classes (as many Marines described them) were not what Marines needed, in discussing his experiences during his deployment to Iraq, Major Madison stated, "The terps [interpreters] were our savior. It took me a year until I could hold a conversation. Some of the customs stuff—don't show the bottom of the feet, right hand versus left hand stuff—wasn't all that helpful."

Similarly, during a field exercise employing cultural training for their upcoming deployment to Iraq, Colonel Redford provided some feedback: "[We need to know] what it is like to truly engage. Not just feet up [don't show the bottom of your feet], but the nuances in building a right relationship."

Over and over I was told that Marines needed to learn material that was *useful* and could be *applied* to the operating environment. Several of the Marines I interviewed made clear distinctions between training that was "touchy-feely" or "behave nice stuff" and operationally relevant culture training. "I don't much care if a Marine is culturally sensitive, I care if he is culturally adept," stated Lieutenant Colonel Gilmore. Similarly, General Edwards commented in a meeting on military training and education, "I'm not looking for cultural sensitivity. I'm looking for cultural awareness. What are we getting out of the training? Is it effective? Is it worth the time invested? Language and culture is a way to save lives. When it fails to do that it loses its usefulness."

These distinctions between cultural "sensitivity" and capability reflect Marines' focus on achieving the mission rather than learning culture for culture's sake. Several Marines I talked to emphasized the importance of learning the "so what"—the operational relevance—of the cultural information they were receiving. "What sells to the guy in the field are the couple of pieces of advice that tie to the field, that change things around," Colonel Irons emphasized. "We need to become relevant to the operator." Similarly, Lieutenant Colonel Lyons, who had just returned from operations in Helmand, Afghanistan, stated, "We had tons of info [on the culture], but I'm just looking at this pile. I need someone to tell me the 'so what' so we can evaluate the information and determine what is relevant."

Perhaps most colorful were the comments of Lieutenant Colonel Jones, who became outraged when the latest in a series of Department of Defense–imposed culture policy directives was sent to his office. The policy document described the latest proposed qualities of a "culturally competent service member," which ranged from "tolerance for ambiguity" to the "ability to use empathy to understand the situation." Tossing the paper on his desk in disgust, he responded, "We're not teaching empathy here!"

The ability to apply the cultural information to the situations Marines would be experiencing was repeated in a number of conversations. "You can sit and talk all day, but eventually you have to get out and do it," stated Gunnery Sergeant (ret) Daly as we discussed the culture training occurring at Camp LeJeune. Similarly, as we were discussing what to put in the culture curriculum, Lieutenant Colonel Gale observed to me, "Teaching a class is one thing, but connecting it to the operation, getting this to be real [that's the challenge]." Continuing our previous conversation, Gunnery Sergeant (ret) Daly explained, "[It's about] the big picture. Job performance. What is the Marine going to do with the information? [You need to] answer the Marine's question. What's in it for me? Why do I have to know this?"

As Lieutenant Adams, just two weeks away from deployment to Iraq, explained, "There are some things that you have to experience to truly understand. You can tell me and tell me. But until I experience it . . . " During a training exercise at 29 Palms, Captain Rhodes stated, "You need to lock onto something tangible, so that Marines can see the effects." And Colonel Irons observed, "If you can't show a Marine that it will be better on the ground from what we're teaching them, then it—[waved his hand over his head, implying 'it goes over their head']. You need that tie to 'this is going to matter.'"

Viewing culture as a skill or tool that can help Marines succeed in their mission, Colonel Thomas observed, "Operational culture has to be tied to mission accomplishment. It's as important as marksmanship."

In the years since CAOCL first began, cultural training has shifted dramatically from "do's and don'ts" to more operationally relevant classes. In describing the cultural materials developed at CAOCL more recently, Major Parker pointed out, "When our product goes out we have to make sure it's meeting the needs of the operating forces."

"We've gone back and managed their [the unit's] expectations and asked, 'What kind of capability do you want on the ground?'" noted Lieutenant Colonel (ret) Charleston in discussing the language and culture training programs at CAOCL. "We showed them how it enhances their capability on the ground and how we can integrate it into their training routines. That's where we really found success."

To do this, CAOCL developed a number of programs that shifted culture and language learning from customs and courtesies to "real-world learning." The center developed a Key Leader Engagement class that placed senior leaders in scenarios where they had to negotiate (usually using a translator) with local leaders from the country to which they were deploying. Instead of classes on Arabic grammar and pleasantries, CAOCL built a tactical language program called Employ Tactical Phrases (ETP 150).[8] The program (which could be translated into any language Marines desired) consisted of a set of classes teaching a list of important memorized words and phrases Marines would need during operations: for example, to use at a checkpoint or searching a vehicle or house. Marines then were required to go through a mock checkpoint or vehicle search with native-language speakers, applying the words they had learned. Most interesting of all, however, were the combined efforts of CAOCL and the leadership at the Marine Corps base in 29 Palms, California, to create a new culminating aspect to their combined arms exercise (CAX) called Mojave Viper.

"Where the Rubber Meets the Road": Learning by Doing at Mojave Viper

A few scrubby bushes line the dry dirt road into Wardah. The desert stretches out along the flat, barren valley, surrounded by steep, rugged, treeless mountains. At noon, as the team of Marines drives into the small Arab village, there is little sign of activity on the blazing hot streets. The houses—square, windowless, cinderblock constructions—seem cheerless and uninviting. A couple of elderly men, robed in white, their heads covered in dark keffiyehs, do not look up as the Marines pass by.

At the town square, the Marines jump out of their Humvees and head into the shade of a long, colorless building that states in Arabic above the door that it is the *makhsan*, the town hall. Inside the *makhsan*, three older men are seated on a rug on the floor, drinking tea. At the front of the team, Colonel Redford steps forward and reaches his hand out. *"Ahlan wa sahlan,"* he says awkwardly in Arabic. One of the taller gentlemen, robed in white and holding a set of prayer beads, reaches out to shake the colonel's hand.

"Greetings to you and your family and the people in this village," Colonel Redford begins. To the right of the colonel stands Emir, a curly, dark-haired man in American clothing who begins to speak quickly.

"Ahlan lianta wa ailitnik wa ilsha'ab fi hatha madina," the translator repeats the colonel's greeting in Arabic to the village elder.

"Wa ahlan lianta wa jundik," replies the town elder. "Greetings to you and your Marines," the translator interprets quickly. *"'Iqad,"* The town elder motions to the rug and waves to his assistant to get more tea. "He would like you to sit down and share a tea with him," Emir explains.

Colonel Redford sits stiffly on the ground, his translator to his side, as two other Marines on his team find spots on the rug nearby. Not waiting for the tea to arrive, he gets down to business. His Marines have been having trouble in Junayniah (a tiny village on the border of Zahar'ra, which has historically been a major center for smuggling), he explains. In the past few months, there have been reports of increased smuggling activity. And they suspect that the smuggling includes guns and explosives.

Quickly Emir translates as the tea is set in front of the village elders. The tall elder listens quietly and unemotionally and slowly sips his tea. He begins to shake his head and replies, Emir translating quickly to the side. The people in Junayniah are not members of his tribe and he cannot help. Yes, there are many bad people there. He is so sorry to be of little assistance.

Colonel Redford's XO (executive officer), Major Abrams, starts to interrupt, when suddenly an immense wailing and crying and shrieking rises up from the street outside. Rising quickly, Major Abrams and his Intel officer go quickly to the window to see what is happening. A procession of black-robed and heavily veiled women are moving down the street, shrieking and tearing their hair out.

"I think it is a funeral, sir," shouts Lance Corporal Binotz, who is standing post at the door. And indeed a few minutes later, a dark car passes by, followed by a number of solemn men carrying weapons.

Colonel Redford turns to the village elder and is about to resume his conversation.

"Boom!" a loud bang explodes out in the plaza. Then suddenly gunshots are ricocheting off the buildings. And screaming—high-pitched, fearful screaming, not the wailing of the grieving women—permeates the air.

In the doorway, Lance Corporal Binotz shouts to the Marines in the *makhsan*, "Sir, an IED [improvised explosive device or bomb]! And gunmen. We're under attack!"

The event described above did not occur in Iraq or any other Middle Eastern country. It was part of a standardized, monthlong training exercise called Enhanced Mojave Viper. The scenario, which took place on the Marine Corps base at 29 Palms, California, spread over hundreds of kilometers throughout the California desert and attempted to reproduce the combat situations of Iraq (and then later, Afghanistan) for Marines just before their deployment. At the time of my field observations, Mojave Viper included several mock "Middle Eastern" villages, two Marine Corps operating bases where the Marine battalions lived and slept in tents and makeshift barracks, and a command and control center (as well as other logistical buildings).

The program began in 2006 and was in permanent operation until the budget cuts of 2013.[9] The focus of Mojave Viper has been to train and prepare Marines for both kinetic (direct fighting) and soft (human) aspects of combat—from "soft skills" such as negotiation with sheikhs to the more traditional "hard combat" skills of fighting insurgents in the street. Describing the cultural training exercises that Marines were required to go through before their deployment to Iraq, Gunnery Sergeant (ret) Daly explained, "Mojave Viper is important because this is where the battalions get real experience— doing vehicle checks, cordon and knocks (an area search),[10] all that kind of stuff—in an urban environment. You can talk about it all you want, but here you get to do it."

Noting that a PowerPoint brief could not possibly provide the same opportunities as a live exercise, Captain Nash compared his training at 29 Palms to their previous work-ups (preparatory training) at Camp Pendleton prior to their arrival, "[During work-ups at Camp Pendleton] I'd give a PowerPoint class and trace out five boxes in the ground in chalk and have Marines practice coming in and out of it. I can do all the IA [Immediate Action] drills and accomplish the motorized ops [vehicle operations] check-in-the box for PTP [predeployment] training. But Mojave Viper is when all of a sudden suddenly I have twenty up armor Humvees and all the assets there."

Echoing this focus on learning by doing, Colonel Irons observed, "Ultimately in the Marine Corps the important measure of the worth of something is in the doing—does it work or not? How do Marines learn? How do they retain—that's by doing things."

Not surprisingly, the Marine Corps' "performance based" approach to training focuses on teaching Marines through doing or practical application. As Gunnery Sergeant (ret) Daly, who had years of experience designing and teaching training programs for the Marine Corps, describes, "Performance-based training—well it's one thing for a Marine who sits in a classroom to regurgitate. It's another to take him into the field and see what he can do."

As a result, most tests of skill level in the Marine Corps are not conducted through written exams. They are based on how well one performs. Reflecting this concern with observable measurable outcomes, Marines have numerous ways of evaluating whether they are successful. Training programs are tied to a list of desired observable skills. Courses and curricula must list learning objectives and measures of effectiveness in their course plans. And field exercises are dissected and evaluated through after action reports and "Tactical Decision Discussion Groups," which provide feedback on the strengths and weaknesses of how well students performed during the exercise.

Central to both the course structure and the observable, measurable performance of the skills in evaluations is the notion that training must be consistent and standardized.

Consistency and Standardization

As discussed in the first part of this book, one of the most important Marine Corps ideals is the notion that there is a consistent set of beliefs, values, and practices that unite all Marines across time and place. Regardless of where one is trained or when one graduated from Officer Candidates School or recruit training, there is a belief that all Marines share the same basic experiences and skills. In discussing his training at TBS (The Basic School) one of my fellow faculty members, Dr. Sorenson (a retired Marine Corps lieutenant colonel), stated, "The experience bonds us together. I was at TBS in the 1960s. But I can sit next to someone who has just graduated and the experience is essentially the same. The curriculum is exactly the same forty years later."

Dr. Sorenson's statement is more of an expression of an ideal of Marine Corps unity than a blanket statement of fact. TBS has certainly made changes in its curriculum over the past fifty years. However, in talking about TBS's curriculum today, one of the former commanding officers there, Lieutenant Colonel Stacy, echoed a similar sentiment regarding consistency in training, "We don't want to make huge rudder changes. We did that in the '90s and that was a disaster. It's important that the curriculum stays consistent across the Corps."

Teamwork and the ability to work together are frequently listed as critical reasons why training must be consistent. "It's all about consistency. Your [Marines] never will know where you're coming from if you're not consistent. If they don't know you or who you are, they're constantly looking over their shoulder. By setting a line of consistency, they know how to respond," stated Captain Rhodes, who was training his company at 29 Palms.

"The important thing about training is that you can go anywhere—Camp LeJeune, 29 Palms, Okinawa—and you learn exactly the same thing," stated

Gunnery Sergeant (ret) Daly. "What we don't want is guys from I MEF working with a platoon from II MEF and they each start off on a different foot. In a crisis we all need to be doing the same thing at the same time in the same way." Given that the Marine Corps is spread out around the world, with Marines constantly rotating in and out of a particular unit, this is a very real concern. The Gunny continued, noting that uniformity in training was necessary so that "if we all come together in Okinawa, Japan, we can all do the same thing."

Staff Sergeant Hagerman, who had recently finished training his unit for deployment to Afghanistan, expressed a similar concern: "This is the military, and we can't have units doing their own thing. We can't have one unit training to A, B, and C and another training to B, C, and D. We need to all be working together."

As both of these staff NCOs illustrate, consistency is essential for Marines to work together quickly and efficiently often under new and rapidly changing conditions. Successful units are expected to establish Tactics, Techniques and Procedures (TTPs): routine methods they all share while working together. These TTPs can range from a standard way to clean the squad bay (barracks) to accepted routines in searching a house in a village in Afghanistan. Sometimes these routines are formalized into written SOPs, or Standard Operating Procedures, which become an official part of a particular unit's documents— a form of memory for an organization where personnel are switching in and out on a monthly basis.

The many challenges of creating an effective working team among rapidly changing personnel were illustrated in two fascinating group interviews conducted one month apart. The first was with a set of platoon commanders who had recently returned from deployment to Iraq and were training for their next deployment with battalion 1/7. The second set of interviews was conducted with the staff sergeants from battalion 1/7 (whose job it was to partner with the platoon commanders). The two interviews provided both the enlisted and officer perspectives on the difficulties these Marines faced in figuring out how to work together with a constantly rotating team of Marines.

LIEUTENANT NANCE: The last deployment . . . we came in, the three of us came in. When did we come in? About two months before we got predeployment leave. He [points to Lieutenant Beyer] came in when we were on predeployment leave. . . . We had—

LIEUTENANT BEYER: Two, three, four . . . three [platoons] with new company commanders, a new battalion commander. So the whole deployment—

LIEUTENANT STAFFORD: Three, the group sergeant major—

LIEUTENANT BEYER: The whole deployment, the whole leadership was brand new three to four months back before we deployed. So we went through that deployment—For a lot of us, as the leaders that was our work up [training] almost. I mean we were on deployment but we were learning so much while we were out there.

The comments of the staff sergeants who were working with these lieutenants reflected a similar sense of chaotic change and the challenges of running a platoon with new leaders:

STAFF SERGEANT BENSON: We get our lieutenants right before the CAX [combined arms exercise] and right before we deploy and then you have to go through . . . it's the same routine with every lieutenant that you get in.

STAFF SERGEANT GONZALEZ: You're going to have a break-in phase. . . . I honestly believe that they come to the fleet and they know they're going to have a problem because they don't know these jobs . . .

STAFF SERGEANT DUKE: I had mine (lieutenant) for two months before we went there . . .

STAFF SERGEANT BENSON: Even our last deployment we got our main dump of lieutenants the week before we went to CAX. Went to CAX, came back, went on leave and deployed right away. So we got here for CAX and a month later we deployed. . . . They didn't have a chance to get known. They were just thrown right in there into it.

Consistency—the knowledge that whichever Marine shows up, he or she is trained to do the same thing—is not only necessary for building an effective team but also for developing Marines who can respond reliably in any situation regardless of the circumstances. One of the advantages of creating a military in which "every Marine is a rifleman" is that they are the *same*—interchangeable, a clearly defined "product." As several Marine leaders explained to me, in a crisis, they always knew they could count on their Marines to do what needed to be done because they had all received the same training. As Captain Nash observed of his company's predeployment training for Iraq, what was important was "training to a standard—knowing that guy can do it there."

"When you look at a Marine, the trousers reach the same point. You know there's a standard that all Marines are set to," said Captain Mills, currently a student at the Expeditionary Warfare School. "I know that there's a certain thing I'm going to get from these guys. You can have personality but education-wise you speak the same language. [It's a] cross-leveling process."

Standardization. Consistency. Performance. Operational relevance. Each of these aspects of training makes sense according to Marines' worldview. The interesting question becomes: what happens when a squishy, "soft," not easily defined or measured concept—culture—becomes something that Marines need to be able to respond to when deployed? While one possible solution is to change the nature of Marine Corps culture and adjust the way that training is conducted, this did not happen. Instead, culture was transformed—processed—into standardized Marine Corps ways of training.

"Marinizing" Culture and Language Training: SATs and the Training and Readiness Manual

As described in the previous sections, CAOCL's ability to translate cultural skills into operationally relevant actions was one approach to creating a culture program that "looked and smelled Marine." However, Colonel (ret) Dallas, the director of CAOCL, believed that without "weaving cultural training into the very fabric of the Marine Corps" such hands-on programs would eventually die. So in 2008, CAOCL began focusing its efforts on creating a culture and language training program that would be standardized, replicable, and applicable to deployments anywhere around the world.

In the Marine Corps, all training courses are written out in a preplanned, programmed sequence, often based upon a formal PowerPoint show called a "brief." Trainers are expected to deliver the brief faithfully, ensuring that every student in each classroom receives essentially the same class in exactly the same manner. The course is based on a set of standards that can then be measured to see if the goals have been met. Gunnery Sergeant (ret) Daly explains, "We have something called the SAT: it's the Systems Approach to Training. The goal is to develop an observable, measurable, performance-based skill set."

Once the training curriculum has been designed according to this standardized SAT sequence, it is then submitted through a formal process for approval as *the* accepted way to train Marines. If the curriculum is approved, it then becomes incorporated into the Training and Readiness (T&R) manual—an enormous manual (now electronic) that contains all of the approved curricula for Marine Corps training. When Marines prepare to deploy, or simply need to conduct annual refresher training, they can download the appropriate classes from the electronic manual and assign someone in the unit to teach it—or request a "trainer" from Ground Training Branch or an associated school.

In the early years of the wars in Iraq and Afghanistan, one of the major critiques that Marines were making about the experts who were hired to teach culture and language courses for the Marine Corps was that they failed to understand the importance of consistency and standardization. Instead, the many different contracting organizations and instructors each had their own particular programs and ideas of what should be taught. In addition, often the instructors would "go off on their own direction" rather than acting as a team player. As a result, the ad hoc, individualistic culture and language programs, which had proliferated during the early years of the war in Iraq, were viewed by many in the Marine Corps as a strange, "non-Marine" approach that did not fit with accepted ways of conducting business. Recognizing the problem, Lieutenant Colonel Jones complained about these programs, "There is a standardization piece. We get smart guys who come and train a unit and it's not standardized, and it's not tied to a training requirement."

In a force in which "every Marine is a rifleman," training must be directed to everyone, not just a select few. Consistency and standardization thus became significantly more important than tailoring the classroom to meet the needs of individual students or units deploying with special needs. The problem was, however, as Major Edwards, a former Military Transition Team member, observed, "there's no real book that says, 'here's a checklist for how to get into a culture.'"

Seeing the importance of making culture and language training fit the accepted Marine Corps processes, over the past years CAOCL has focused upon creating a standardized curriculum for the entire Marine Corps. This curriculum is based on an approved format—a template—of cultural aspects that Marines should know about any country to which they deploy. The template then can be filled in with the specific details for any country or region where Marines are headed, making it consistent and standardized and suited to a Marine Corps that deploys anywhere at a moment's notice.

After much debate about the numerous acronyms and templates that were being used by various services to categorize culture (for example, ASCOPE, PMESII, DIME),[11] the Marine Corps settled on the "Five Dimensions of Operational Culture," which had been outlined in a handbook designed initially for the Marine Corps University classroom.[12] As the coauthor of this publication, I was most surprised to discover that the five basic anthropological concepts that we had identified to help Marines think about cultures (their use of the environment; exchange and economy; social organization; political organization; and belief systems) had suddenly become quasi-doctrine and the foundation for a standardized one-size-fits-all approach to training Marines about culture!

Using the Marine Corps SAT method, these Five Dimensions became the template for an official standardized culture training package that could be downloaded by any Marine, anywhere in the world, and used to train his or her unit. The course would be accessible online, just like any other training class, whether it was learning to shoot a rifle, disarming an IED, repairing a motor, or preparing to interact with a foreign culture.[13]

Major Parker, who participated in the development of this program at CAOCL, explains, "[We're working on] service level standardization. Making sure that what a Marine gets on the East Coast is the same as what he gets on the West Coast. . . . So we are formalizing [culture] training tasks which are tied to the Training and Readiness manual.[14] [That way] when your boss requires training it's tied to institutionalized standards."

Although the resulting program lost the flexibility and cultural complexity of earlier efforts, as two captains discussing their culture training observed, it met the needs of the Marine Corps better than the early individually tailored programs for Iraq. "In previous deployments I would say that company commanders and battalion commanders could really do whatever they wanted to effectively train the battalion," stated Captain Rhodes. "Now with the PTP [predeployment training] message coming out, it's very directive and specific of what you're gonna do. And if you're gonna adhere to that and follow that order religiously that's going to take absolutely all of your dwell time [time in the United States] just following that. So effectively—good or bad, however you want to look at it—your training is more or less written for you, it's just your job to execute that training and follow through. "

He continued, "I think it's a good thing because the problem with having the freedom to act in a very specific environment where everyone is going to combat, more or less, the battalions become more personality driven than actual T&R manual driven, you know."

"And it standardizes everything," Captain Neal added. "So the battalions that have done the PTP program religiously and done Mojave Viper and gone over there are far better prepared, and more importantly they are more standardized in their outlook and their training over there."

CAOCL's past and continued survival, particularly in light of the many start-up organizations in the Marine Corps that have already disappeared in the past few years, provides a critical case study of the importance of adapting external governmental policies to fit within the culture and ideals of the organization. In the case of culture and language training for the Marine Corps, this has been accomplished by creating programs that are operationally relevant and focused on "learning by doing," and by standardizing the training so that it is accessible to all Marines deploying anywhere in the world.

The end result has been a culture and language program that fits Marine Corps ways of doing, thinking, and interacting with the world. Rather than changing the Marine Corps to adapt to a new way of teaching and training these unfamiliar concepts, the theoretical cultural concepts—the actual way that people in another country view the world and interact with each other—become "*processed*" into something familiar to Marines. Culture, then, is restructured to fit into the Marine Corps way of conducting business. As a result of this "processing," culture becomes less foreign and strange, and thus less likely to be rejected by those in the organization. Culture becomes, as I discuss in the next chapter, *reshaped* into something that "looks, smells, and tastes Marine."

9 "There's No 'I' in Team"

Reshaping Culture Specialists to Fit a Nonspecialist Culture

We need to change manpower assignment policies to keep experts in a region/language working in that region, otherwise your training burden will never decrease. Sending Korean linguists to Iraq is a waste of time, but that is what was done in my unit in 2004.

—*Forty-two-year-old Marine Lieutenant Colonel and Communications Officer (CAOCL survey)*

The USMC FAO [foreign area officer] program is a "career ender" for what otherwise might be highly qualified and motivated officers [who develop special language and culture skills]. They are also under utilized in their target region [the cultural region they have studied] with a rate of utilization hovering around 26% according to 2005–2009 placement. Think about it . . . if a pilot left flight school then had a 26% chance of flying an aircraft that he spent more than two years studying, this would be a SERIOUS manpower underutilization problem. Currently, USMC FAOs spend one year at NPS [Naval Postgraduate School] on a general security studies master's, then 6 to 12 months in DLI [Defense Language Institute] and most, NOT ALL, complete a one-year ICT [in-country training] in their target language region. Then, the officer returns to the fleet [Corps] with three years of unobserved fitness reports [performance evaluations] and preparing for the next promotion board . . . this needs to be addressed.

—*Thirty-two-year-old Marine Captain and FAO (CAOCL survey)*

OVER THE SIX YEARS THAT I HAVE FOLLOWED THE Marine Corps culture venture, Marines have not only begun to recognize the importance of culture but also to apply cultural knowledge and skills in increasingly creative and sophisticated ways. However, as the captain's comments above reveal, institutionally the Corps has been reluctant to develop permanent regional and culturally focused programs or invest in individuals with specialized cultural and language skills.

As a result, while individual Marines may recognize the importance of cultural skills, from an organizational perspective, the institution of the Marine Corps has been reluctant to adapt its structure or shift personnel assignments so that Marines can be placed in billets where these cultural skills can be used effectively. As I explain in this chapter, this is not simply due to organizational inertia or insufficient guidance from leadership. The problem is much more fundamental. For such shifts challenge the very essence of Marine Corps identity: a Corps based on the philosophy that "every Marine is a rifleman." If the foundation of the Corps is a dedicated, flexible team of Marines who can do anything anywhere—the expeditionary culture of Lance Corporal Binotz and "semper gumby"—can and/or should the Corps shift its assignments and teams to include niche Marines with specialized culture and language skills?

For the Marine Corps, the answer has been "yes and no." Rather than develop specific culturally oriented programs or identify and deploy Marines with particular regional and cultural knowledge, the Corps has adopted a truly Marine solution to institutionalizing cultural skills. Today's Marines continue to be sent anywhere around the world, regardless of (and perhaps despite) their personal regional or cultural backgrounds. However, in a fascinating *reshaping* of the Department of Defense (DoD) policy imperatives for cultural specialists, the Corps has developed two unique Marine-focused solutions to provide a general overall cultural capacity to these expeditionary units: the Regional, Culture, and Language Familiarization (RCLF) program and the creation of several new military occupational specialties (MOSs) that do not focus specifically on cultural or language skills but do require basic cultural competency in their training.

"Every Marine a Rifleman": The Marine Corps' Manpower System and the Problem of Specialists

Ten years after the invasion of Iraq, the Marines in my study and in my classes are a different war-seasoned generation, with one, two, or even six or seven tours in Iraq, Afghanistan, and other countries around the world. Marines no longer talk about basic cultural information: what a tribe is, how to conduct a Key Leader Engagement, or whether cultural narratives are important. Instead, my classes and emails are filled with nuanced discussions of the tribal and ethnic composition of Helmand province, the challenges of negotiating with religious imams or using interpreters, and even culturally insightful stories about their use of Afghan poetry in information operations!

Despite these shifts in attitudes among individual Marines, however, the Marine Corps manpower and personnel assignment system has changed little, if at all, in response to the cultural aspects and requirements needed for deployments and assignments to foreign countries. Rather than selecting specialized Marines who have specific cultural or language skills for a particular mission or deployment, units continue to be built and deployed based on a standardized set of needed billets (referred to as "spaces" in Marine Corps parlance) matched to the personnel available (referred to as "faces"). Heritage-language speakers are often not identified or deployed to places where their language is spoken. And foreign area officers (FAOs) and regional affairs officers (RAOs) are assigned, on average, only 25 percent of the time to the region they have studied.

In fact, the Corps has fought vehemently against the assignment of any "special" group of Marines whose unique skills could be tailored to the culture and language of a particular region of the world. Even when required by DoD to provide a cadre of Marines to support Special Operations Command (SOCOM), which sends out small, mobile, specially trained groups that must engage with the local population, the Marine Corps resisted until finally it was forced to comply.

This resistance to specialized assignments is not because the Marine Corps does not recognize that language and cultural skills matter. It is because the Marine Corps' organizational culture, and its resulting manpower and personnel assignment system, is focused on creating and deploying a flexible, generic "everymarine"—the complete and exact opposite of the specialized expert that DoD policy experts are demanding of the U.S. military services.

Marine Heritage-Language Speakers

Perhaps one of the greatest ironies in a Marine Corps culture that is "color blind"—accepting all Marines as equal regardless of individual characteristics such as race, gender, or ethnic background—is that, as a result, individual cultural and linguistic differences between Marines are overlooked and even ignored in order to create an interchangeable set of equally qualified, all-purpose Marines. Because "all Marines are green," the Marine Corps does not systematically collect data on the ethnic, linguistic, or cultural backgrounds of its members or otherwise emphasize the inherent differences and unique qualifications of its Marines.

Correspondingly—reflecting and reproducing these "every Marine a rifleman" ideals of the Corps—over the past ten years the Marine Corps Manpower and Reserve Affairs (M&RA) department has done little to change the way it assigns available individuals (faces) to billets (spaces). Personnel assignments continue to be made according to a standard list of open billets designated by MOS. Unless a command deliberately identifies a clear need for an individual with specific cultural or linguistic skills, Marines are rarely selected or sent to regions where their multicultural backgrounds and language skills could be of advantage to the mission.

Further compounding the problem, while individuals come to the Corps from many different cultural backgrounds, the organization has no official way of capitalizing on their varied language and cultural skills unless the Marine specifically offers the information. Although the Marine Corps does offer incentive pay (Foreign Language Proficiency Pay, called FLPP) for Marines who speak a foreign language and test successfully on the Defense Language Proficiency Test (DLPT), Marines must volunteer for the test and pass it.

How significant is this deliberate bias against recognizing (or rewarding) individual difference in the Marine Corps? Considering that over one-third of the Marines who responded to the Center for Advanced Operational Culture Learning (CAOCL) survey responded that they had useful prior language and culture skills, the bias against identifying these skills comes at an enormous cost to the Corps. Thirty-three percent of the respondents to the survey stated that they "speak, read or write a foreign language," and 38 percent said they "grew up in a multicultural environment within the U.S. or overseas." Even more telling, 42 percent of the enlisted Marines reported that they "grew up in a multicultural environment" in contrast to only 25 percent of the officers. (These differences were statistically significant $p < .000$).[1] Given that enlisted

Marines are more likely to come from minority and ethnically diverse communities, this difference is not surprising.

In the CAOCL survey and in my interviews, a number of Marines expressed great frustration with the Corps' inability to leverage the language and cultural knowledge of its own members. One Marine sergeant and automotive technician stated on the CAOCL survey, "In the case of the earthquake in Haiti on Jan 12, 2010, there were many Haitian Marines aboard Camp Pendleton, CA who speak Haitian Creole and French who could have been a lot of help out in country translating for the US troops and UN troops. However, the Marine Corps never used them." Another survey respondent, a Marine captain and infantry officer, added, "We have the assets in the Marine Corps already. We are terrible at identifying them. I work at MCRC [Marine Corps Recruiting Command] and proficiency in foreign languages is not even asked by recruiters as kids join. We have kids joining today fluent in Russian getting assigned to be bulk fuel specialists."

In discussing his deployment to Iraq, Major Neal narrated a story that illustrated the way that his unit accidentally discovered the language skills of one of its Marines while deployed in Iraq:

> We had a mailroom and one day I went on in. I saw the name on a young enlisted Marine's shirt. Because I had lived in Egypt [and recognized it was an Egyptian name], I asked, "Are you Egyptian?"
>
> He said, "Yes."
>
> I asked, "Do you speak Arabic?"
>
> He said, "Of course."
>
> So I said, "What the hell are you doing in the mailroom?"
>
> Our unit had no one who spoke Arabic well. I spoke it only poorly. We had an Iraqi who spoke Arabic and German, and then someone who spoke English and German. You can imagine. . . . So we pulled this guy from the mailroom, and he was very helpful in our mission.

From an outsider's perspective, it would seem obvious that there is a simple solution to this problem: create a mandatory Marine Corps database of the language and cultural skills of every Marine. However, many Marines deliberately hide these skills from the Corps, specifically because they want to maintain some stability and control over their personal and family lives. Puzzled about Marines' reluctance to report their language skills, I spoke to Colonel Irons, who explained the problem to me:

The underlying precept to the manpower system is the needs of the Marine Corps. If we can meet the needs of the individual and his desires and wants, we will look for that. But at the end of the day, it's the needs of the Marine Corps that determines outcome.... So if the guy is a Chinese speaker and they need that, they can pull him [out of his current position and send him elsewhere].

For example, if we know we're going into Haiti, they'll look in the database and they'll take them [Creole speakers]. If someone knows you have a skill and they need it, it's the needs of the Marine Corps that trump everything. That's why so many Marines don't self-declare [their language and other skills]. You have no status. You can argue all you want. You can plead your case. But you have no status. The Marine Corps owns you. If you refuse your orders [to go] you'll be out in six months after your assignment....

People just don't want the Marine Corps to know what skills they have. Nobody wants to be jerked out of your situation, especially if you are just coming back from deployment.

Not only is there no comprehensive database reflecting the cultural and language skills of Marines, but even if the Corps did have one, due to the way that the manpower system is designed to assign Marines to needed billets, it is often difficult to match specialized Marines to locations where their skills could be used. Continuing with my interview with Colonel Irons, he described the current method for assigning the approximately 200,000 active duty Marines in the Corps:

It's a mathematical problem. We talk about "spaces" or billets and "faces"—the Marines who fill those billets. There isn't a Marine for every billet. Not a one for one match. [So we have] a priority list: "Accepted" get 100% of the people matched for the MOS. "Priority" gets about 90–95%. "Proshare" gets everything left over.

It's fairly complicated. I'm simplifying. But basically you put it all into a large database. You have the billets, the command priority, and your inventory [of available Marines].... You push a button, do the math, and out spits the enlisted staffing model and the officer staffing model.... In typical military decision-making fashion, we don't tend to take consideration of the individual. There is a little more latitude with the officers. But the processes are not refined to do that. There isn't the time and money and personnel. But also what's the driving factor? It's the needs of the Marine Corps. There's not a lot of tailoring especially at that first enlisted level.

As a result, several of the Marines in the survey observed that due to their military occupational specialty or the billet to which they had been assigned, they were unable to use their language and cultural skills despite repeated efforts to apply them to the Corps' advantage. For example, a Marine judge advocate expressed his frustration when he noted on the survey, "The Marine Corps also under-utilizes the skills it already has available. For example, I speak Farsi and can understand Dari well, languages spoken in and very useful in Afghanistan. However, I had tremendous difficulty deploying because of my MOS and duty station." Similarly, a Marine captain suggested, "Give Marines who are fluent in critical languages more of an opportunity to use their language. I speak fluent French and it's on my MOL [Marine online database—which contains personnel records], but I've never been approached or offered an opportunity to use it. I would love to be able to use my language skills sometime, but I really don't know of any billets in the Marine Corps or joint billets where I can use it or that I can apply for."

Recognizing these difficulties, in a recent new publication released by the Corps, the "Marine Corps Language, Regional and Culture Strategy: 2011–15," the authors specifically mention the need to identify heritage speakers in the Corps and to better track and place them in appropriate billets.[2]

Yet not all Marines viewed the inability of the Corps to adjust its personnel assignments as problematic. As Major Swanson, an instructor at Officer Candidates School, noted, there is a logic to the Marine Corps manpower assignment system. Given the Corps' cultural emphasis on flexibility and adaptability to each new mission, he argued that selecting Marines to undertake specific roles based on specialized skills would be to the overall detriment of any unit. "In the Marine Corps you can't mess with the sanctity of 'every Marine's a rifleman.' Otherwise you're bringing in someone for Admin [administration], someone for a translator. If you have a Marine who's an interpreter, he's not standing post [doing guard duty]. He's not cleaning up. It won't work."

Foreign Area Officers and Regional Affairs Officers

The challenge of identifying and utilizing language or cultural specialists in the Corps is not limited simply to Marines with heritage-language and cultural skills, however. The issue is most pronounced with those Marines who are specifically prepared to provide these skills to the Marine Corps: foreign area officers (FAOs) and regional affairs officers (RAOs). As with similar programs in the other U.S. military services, the Marine Corps invests an immense amount of time and resources to educate and develop FAOs and RAOs. Both FAOs and RAOs are paid to complete a master's degree focusing on a region of the world. FAOs continue on to receive a year or more of intense language training and then often spend another year traveling in the area they have studied.

However, unlike in the other U.S military services, FAOs and RAOs are not primary MOSs (meaning the main job of the Marine). Thus Marines who do study a language and region through the FAO and RAO programs must still maintain their skills in their primary military specialty to be promoted. As a result, it is not uncommon that after studying a region for several years, a Marine FAO will be assigned to a tank battalion or an administrative position far from the region he or she has studied. Not surprisingly, the language and cultural skills developed at a tremendous economic cost[3] atrophy soon after a Marine's formal education is over. And because of the almost impossible competing demands of performing well in their primary MOS while working in billets as a FAO, most FAOs do not advance in their careers beyond the level of major or lieutenant colonel.

Interestingly, although the CAOCL survey asked only one question regarding FAOs and RAOs ("Are you a FAO or RAO?"), several respondents took extra time at the end of the survey to comment on their experiences. One thirty-six-year-old Marine major and FAO was quite vocal in his response:

> The USMC FAO program is absolute SHIT! The Marine Corps devotes significant resources to the FAO training pipeline, but virtually nothing after that. A couple of years ago they invented a selection board process for Marine attachés that favors FAOs, much more likely a reaction to a DoD requirement than any show of initiative. The FAO program needs two things. It needs a program for sustainment (relatively inexpensive, maybe $10K per officer per year) and it needs billets better matched to the USMC's current FAO inventory. The current "FAO-coded billet" list was concocted by an amateur with his head up his ass. Take an annual look at the list of FAOs in the force, determine priorities, identify opportunities and be flexible. Right now the FAO program's benefactors are not thinking broadly or with much flexibility—killing and wasting the entire program in the process.

As the major notes, one of the greatest problems for FAOs is that they are often not matched to billets that are relevant to their skills and education. Yet even if there is a good match, it is often difficult for FAOs to use their skills effectively. As Major Neal, another FAO, commented, "I have never spoken to a commander who knew what a FAO is supposed to do for him. It is not written down in any manual."

As with the Marine heritage-language speakers, one of the main reasons cited for the Marine Corps' limited use of FAOs is the Corps' cultural emphasis on creating a general purpose force: "every Marine a rifleman." As Colonel Irons observed, "If you look at the fact that the Marine Corps is the only service that hasn't broken out the FAO as its own career track, that speaks to we're not specialists. Even the Navy—that probably has the least contact with the indigenous population—has a separate career track for FAOs. . . . We've really discounted them. We don't promote them. We don't encourage them to do multiple tours [in the region they've studied]. We encourage them to get in, get out, and get back to the fleet [Marine Corps general operating forces—usually deployed]."

Until recently, only Marine officers could become FAOs—an interesting limitation given the high percentage of enlisted Marines who already come to the Corps with a multicultural background. However, this is changing. Recognizing the need for senior staff non-commissioned officers (SNCOs) with cultural and language expertise, the Corps has begun a new parallel foreign area officer program for SNCOs who can now become foreign area staff non-commissioned officers (FASs). [4]

Although it is tempting to blame the Marine Corps Manpower and Reserve Affairs department for its inflexible approach to assigning linguistically and culturally specialized Marines to appropriate billets tailored to their specialization, my interviews suggest that the issue runs much deeper. Ultimately, all Marine Corps organizations must reflect and reproduce the dominant cultural values of their service; otherwise, they live a short and poorly funded life. The Marine Corps manpower assignment system is no exception. It is designed to fill personnel rapidly and efficiently into units that may deploy at a moment's notice. Devoting extra resources to identifying individuals with unique capabilities and then pulling them away from their current billets is time consuming and causes additional gaps to fill due to the specialist's absence. As Colonel Irons explained, "It comes down to manpower. The guys that fill this thing have to come from somewhere." The current manpower structure, then, is designed to give every unit a Marine rifleman—not to create specialized teams of outstanding stars.

"All Marines Are Special": The Battle over MARSOC

The Marine Corps' opposition to developing and assigning specialists rather than team players is illustrated in particular by the battle over the requirement by DoD to provide Marines to SOCOM (Special Operations Command) to form MARSOC (Marine Special Operations Command). All of the U.S. services were expected to contribute members to this elite joint group, whose function is to perform dangerous assignments among the populations of foreign countries in small teams reminiscent of the Army's Special Forces. According to the MARSOC website:

> MARSOC Marines must be mature, intelligent, mentally agile, determined, ethical, physically fit and able to contribute to and collaborate as part of an independent team. They must also be able to function in remote, ambiguous and complex environments with limited support.
>
> So if you want to . . .
> • Travel to remote regions and work with different cultures . . .
> • Become eligible to participate in highly specialized training . . .
> • Be an integral part of a small, close-knit team . . .
> • Work in a dynamic, fast-paced environment . . .
> . . . *take the challenge.*[5]

Given that Marines love a challenge, and that the Marine Corps ethos emphasizes bravery and a Spartan rugged identity, one might have thought that the Corps would have been the first to volunteer its men and women to SOCOM. However, instead, viewing SOCOM as an elitist program that would single out certain Marines as "special" (even the title was problematic), the Marine Corps fought the initiative. "We have fought MARSOC for years because we didn't want to create a special group," Lieutenant Colonel Jones observed, illustrating how important an egalitarian ethos was to the Corps. "I think we gave in for political reasons."

Colonel Thompson echoed this concern with creating an elitist special group: "MARSOC was a gun to our head. [We didn't want] to stand MARSOC up but Rumsfeld wanted SOF [special operations forces] capability to contribute to SOCOM. MARSOC doesn't belong to us it belongs to USSOCOM [U.S. Special Operations Command]." Then he shook his head and added, almost spitting out the word "special": "Special Operations. We used to shoot special people! The Commandant said 'all Marines are special.' For the past ten years Marines have been doing these things." Colonel Thompson paused, to give his own thoughts on the notion of *special* forces, "SOF is not special people but terrific people with special training that develops special skills."

General Midway explains the origin of this resistance to the creation of special Marine units: "Since World War II, there has always been an aversion to this creation of a unique unit. In World War II, the raider units were viewed as elite. Then the decision was made that this is separating Marines by unit."

Interestingly, although the Marine Corps did ultimately comply with the demands of DoD and stand up (create) its own special operations force in 2006, many Marines still do not see the program as being part of the Corps. "MARSOC guys—that's kinda squishy," stated Colonel Irons. "I would tell you we don't have MARSOC guys because we are a general purpose force."

This cultural bias against creating a force of specialized Marines underlies and influences the way that the Marine Corps is structured. Reinforcing and reproducing this nonspecialist organizational structure is the task of the Marine Corps M&RA. A number of Marines that I interviewed expressed frustration with the inflexibility of the current manpower system, attributing its inability to adapt to the new irregular warfare environment to an outdated "Cold War mentality." As Captain Nash noted of the challenges of personnel assignments in the company he was commanding, "Our manpower system is not equipped [to handle the current operational demands], and I don't think you'll find a guy in this Marine Corps that disagrees with me on this. Our manpower system was designed in the '70s for a certain type of operational environment. The operational environment's changed, and we've put a lot of patches on it to make it work because we're Marines. But the system still hasn't quite regulated, and it's almost like we don't want to completely change it, because we think, 'what happens when we switch . . . switch to the next thing?'"

Captain Nash's comments reveal a primary concern that many Marines, including the past two commandants, have expressed regarding the creation and assignment of specialist Marines with niche capabilities: once the United States moves on from the current counterinsurgency-focused conflicts in Afghanistan and Iraq, what will the "next thing" be? And will this new "post-COIN" conflict environment still require Marines to have specialized cultural, regional, and language skills in order to succeed in their missions?

"Semper Gumby" Marines: "Culture General"
Skills versus Regional and Language Specialists

In recent years, numerous documents and reports by various Marine Corps organizations have been written about the future operations of the Corps.[6] Closely tied to this debate has been the question of how the Marine Corps should be structured to fit these future operations. A report from the 2010 Marine Corps Force Structure Review Group recommended a future structure that continues to emphasize the flexible, expeditionary nature of the Corps and that "provides a force optimized for forward presence . . . [and] provides readiness for immediate deployment and employment."[7] The report also included a mention of cultural capabilities, stating that key actions to be taken should include "retaining and better integrating the training, advising and assistance organizations designed to enable and enhance irregular warfare capabilities, partner-nation engagement and cultural understanding."[8]

According to Lieutenant Colonel Simons, who was working at Marine Corps headquarters, in response to the anticipated force structure and skill needs of the Marine Corps, in the coming years the Corps anticipates shifting its culture and language efforts away from specialized predeployment culture training. "As we look past OEF [Afghanistan] the question is how do we prepare for withdrawal from different areas? We're going to keep the PME [professional military education] piece but when we want to know something specific we'll need to tap into outside experts. . . . We can't have [Marine language and culture] experts in every building."

As Lieutenant Colonel Simons's comments indicate, the emphasis in the future will be on educating a Corps of generalists—not specialists. If specific language or cultural expertise is required, the assumption is that the Corps will reach outside of its Marines to locate that expert. Thus, rather than fundamentally changing the structure or ideals of the Marine Corps to accommodate to the need for cultural and linguistic specialists, the Corps will maintain its own "every Marine a rifleman" generalist culture.

At first glance, it would appear that by moving away from programs that develop culture and language specialists, the Marine Corps is directly contradicting DoD requirements to build a sustained culture and language capability among its military members. However, from the Marine Corps' perspective, this is not the case. Instead, the Corps has reinterpreted these external policy directives—*reshaping* them to fit within the Marine Corps way of conducting business.

To do this, the Marine Corps has focused on the development of what internal Marine policy documents refer to as "culture general" skills: creating generic, cross-culturally competent Marines who can operate in many diverse areas of the world, instead of building more experts with regional or language-specific skills. This "culture general" education, according to the recent Marine policy document "Marine Corps Regional, Culture, and Language Familiarization Program," "is a conceptual approach that guides Marines in framing the problem and then seeking the appropriate culturally specific information to solve the issue. Culture general concepts can be applied to any culture around the world and are therefore applicable to any operational environment."[9] The document specifically contrasts these "culture general" skills to language and regional skills, which it defines as "culture specific education/training [that] pertains to the unique characteristics that are attributable to the people of a particular geographic area."[10] The Marine Corps' distinction between these skill sets—"culture general," "culture specific," and "language skills"—is represented in a diagram published in the document (see Figure 9.1).

To develop this "culture general" capacity, the Marine Corps has come up with two clearly Marine-specific solutions. First, the Corps has added several new MOSs and programs (such as advisor training and civil affairs), each focusing on developing general skills in working with the local population. Marines with these new MOSs receive greater education and training in general cultural aspects of operations. Second, the Corps has created a rather paradoxical program—RCLF—a regionally focused program that is based on the assumption that Marines may never actually be deployed to the location they are required to study throughout their military career.

MOSs Requiring Cultural and Language Skills

Sitting in my office, dressed impeccably in a black, pin-striped suit and stiff white shirt, Master Gunnery Sergeant Rojas is the perfect gentleman—highly educated, refined, well spoken, and a native Spanish-speaking enlisted Marine. "I was born in Colombia," he begins the interview. "My family moved here when I was two, and we lived in New York City before relocating to Miami. In my home, my family forced us to speak Spanish. . . . Immediately after graduating high school I joined the Marine Corps. I served four years active duty and left as a sergeant. I then went off to college and received a degree in poli-sci."

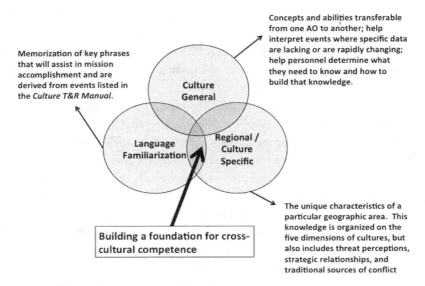

Concepts and abilities transferable from one AO to another; help interpret events where specific data are lacking or are rapidly changing; help personnel determine what they need to know and how to build that knowledge.

Memorization of key phrases that will assist in mission accomplishment and are derived from events listed in the *Culture T&R Manual*.

Culture General

Language Familiarization

Regional / Culture Specific

Building a foundation for cross-cultural competence

The unique characteristics of a particular geographic area. This knowledge is organized on the five dimensions of cultures, but also includes threat perceptions, strategic relationships, and traditional sources of conflict

FIGURE 9.1 Building a Foundation for Cross-Cultural Competence in the Marine Corps. Source: US Marine Corps Center for Advanced Operational Culture Learning

We have been discussing how the master gunnery sergeant had ended up leading Marine military advisor training groups in Latin America. "My tours and experience started in the '90s. I was recalled to active duty and sent to Honduras as a linguist for SOUTHCOM [southern military command area corresponding to Central and South America).[11] After my tour in Honduras, I was sent to Puerto Rico as an Intel [intelligence] analyst, then to Colombia and Paraguay."

"I jumped out of the active reserves to attend law school." He paused and then continued, explaining how his commitment to the Marine Corps and his country was greater than his desire for personal wealth through a legal career as a private lawyer. "But instead of practicing law, I was back on orders with the Joint Forces JTF-Six/ JTF North][which corresponds to North America] as an Intel analyst and linguist working counter drugs and border security, with the U.S. Border Patrol, DEA [Drug Enforcement Agency], FBI , and other federal law enforcement agencies. It was all this experience that eventually brought me to MCTAG [Marine Corps Training and Advisory Group].

"MCTAG started out as the Marine Corps special operations group but it got renamed. Eventually it became MCSCG [Marine Corps Security

Cooperation Group]. . . . I was in charge of getting the Marine trainer advisors ready for MTT[military training team] missions into our AO [Area of Operations] within Central and South America. Knowing both the culture and language made me the senior trainer advisor of the teams, and for close to three years we conducted successful security cooperation MTTs.

"Because I was fitting a quasi diplomat role, [understanding] culture was very important. Every time we came back from our MTTs I would bring back relevant information to update the language and culture training. During the MTT missions I would observe my teams and take [important operational] phrases, record them, and bring them back. [We went everywhere]: Panama, Nicaragua, El Salvador, Honduras, Guatemala, Ecuador, Colombia, and Peru; all throughout Central and South America.

"The cultural info was so important. We were instructing everything from land nav [land navigation], patrolling, infantry skills, etc. . . . but they [the Latin American militaries] would always ask for more training. We were there to empower their personnel, give them the skills to better control the security of their borders, and counter the existing transnational criminal threat. So they wanted every type of training we could provide."

Advisor training, Master Gunnery Sergeant Rojas noted, was far from a risk-free assignment. "Our teams would go in as two-to-seven-man teams. We would sometimes go in unarmed. You really did need to know [the local] culture and language in order to get yourself out of potentially dangerous situations. There were days we were going to the training locations and we'd see [dead] bodies on the road."

What were some of the challenges of working in a new advisor training program, I asked, scribbling as quickly as I could in my notebook.

"The problem was, I was losing all these great SMEs [subject matter experts], linguists, and trainer advisors when their reserve orders were up. The initial trainer advisor teams were all reservist. The active duty GPF [general purpose forces] were all focused on Afghanistan, so the reservist would be called up to man these teams for the SOUTHCOM AO [South America Area of Operations]. . . . The high turnover of trainer advisors was unacceptable. They would come on board [join] for a year. Some would opt to continue for up to two years. The op tempo [operational tempo] for the teams was sometimes very high, and we'd have Marines jumping on and off the teams.

"One of the benefits—the reservist brings something to the program that the active duty component has to learn. The Marine Corps military culture doesn't work well in these Latin American countries. . . . There's a difference in social skills between reservist and active duty Marines. The active duty Marine is Marine," he sat up rigidly, arms ramrod straight down at his side to illustrate. "The reservist is more laid back, compassionate, flexible, and patient. So that makes him better qualified to successfully interact with Latin American counterparts."

He explained, "If you are a Marine and your host nation counterparts fail to show up for training at 0500, you don't dog them out [like you would in the Corps to a fellow Marine]. We learned to ask questions first: 'Why are you late?' That's not normal for Marines to ask or care. But then we would discover . . . "

Master Gunnery Sergeant Rojas began imitating his Latin American military counterpart: "'I had to get up at 0300 to get here on my own dime. I live three hours away and had to take the bus; my service doesn't pay for this transportation.' These scenarios would give our trainer advisors new perspectives on the host nation counterpart's realities and how different their circumstances were from country to country.

"The social skills that reservists brought—firefighters, police officers, lawyers like myself—gave us a more keen sense of understanding and patience when working with our Latin American counterparts. We had more to talk about than military or Marine talk. Most of us had families. We'd get there and talk about our families, kids, civilian jobs. We could bond more on a social level as opposed to an active duty guy who would like to talk about Marine stuff—training, weapons, combat. They can't go down [there] with the traditional Marine Corps warrior mentality."

Despite the challenges, Master Gunnery Sergeant Rojas was hopeful about the program and its future. "Security cooperation works. We would come back and see it in our newspapers: 'This amount of drugs was seized [in Nicaragua, Panama, Honduras].' And you'd point to the guys [in the article]: We trained those guys, or know them.

"So the Marine Corps is now moving [this security cooperation program] more to a general purpose force. They are acknowledging that it works. MCTAG has now turned into MCSCG–Marine Corps Security Cooperation Group. There's even talk about designating a Trainer Advisor MOS. Trainer Advisor Marines have to be more culturally sensitive, know the history, psychology, and current situation of their Latin American counterpart. You have to know much more than infantry skills."

Master Gunnery Sergeant Rojas and I have been speaking for an hour and a half, far longer than I had promised when I requested my interview. So I began to close up my notebook and thank him. Then, as often happens when I have finished my interviews, there is that second conversation, that post-interview, that sudden unexpected gem of a comment that brings the entire interview into sharp focus.

Looking at me firmly, as he began to stand up, the Master Gunnery Sergeant added, "I have seen the change in the Marine Corps—not just internally, but the multicultural quality of the people coming in. The positive by-product coming from the lessons learned in Afghanistan and Iraq is that we are now forced to take a different approach—with more emphasis on culture and language. We were forced to learn. The Marine Corps is now using their personnel as cultural resources. I could be practicing law, yet here I am still working for the Marine Corps."

In the past few years the Marine Corps has created and expanded a number of new programs and MOSs that focus on the human aspects of the conflict environment, such as the advisor training program described by Master Gunnery Sergeant Rojas. The very nature of these new programs and military occupational specialties, which include advisor training, civil affairs, information operations, psychological operations, and culturally focused intelligence MOSs, requires that Marines with these military specialties must build skills in interacting with the local population—thus creating a focused "capability" (as Marines refer to it) in culture and language skills.

As in the case of the advisor training program (which changed its name, structure, and location several times over a six-year period), and reflecting the Marine Corps "semper gumby" approach to achieving the mission that characterized early operations in Iraq, many of these new programs have been in constant flux, manned to a large degree by ad hoc staffing and supplemental forces such as reservists. (Marine reservists must complete recruit training or OCS just like the rest of the force. But then they go on to live a civilian life punctuated by monthly military training, unless called up on "active duty" for a specific period of time.)

Advisor training groups are not the only ones that are predominantly composed of reservists and other ad hoc members. Like military advisors and FAOs, civil affairs Marines (who help provide stability to an area by building infrastructure such as roads, bridges, wells, and schools) do not hold a primary MOS in the Corps. Equally challenging for civil affairs programs is that unlike the Army, which can afford to have a permanent civil affairs corps, as Dr. Green explained to me, "all civil affairs assets in the Marine Corps are reserve."

According to Major Perkins, an instructor at the Expeditionary Warfare School, this reliance on supplemental and reserve forces is a typical solution for providing specific expertise to the Corps. "Our Marine Corps training is not MOS centric. We're capable of veering off into anything. Our culture supports that. We prefer to have someone who is technically inclined to do certain things, but anyone can [be asked to] do it. Reserve officers often fill billets we have no training for." According to Major (ret) Bates, who was now working as a contractor, the preference to avoid specialization means that the Corps often does not have permanent Marines with the necessary skills in certain MOSs: "The emphasis on homogeneity means the Marine Corps doesn't tend to retain Marines with special skills. We have a limited capacity in civil affairs, engineering."

The Marines' flexible approach makes it possible to quickly create new programs and adapt to new challenges. However, the reliance on supplemental forces also has its disadvantages. As Gunny Rojas observed in his work with the advisor teams, these supplemental forces are constantly in flux, often staying for a year or less. Major Edwards, who had deployed on two military transition teams and one police transition team in the past four years, shared similar concerns about the nature of such transitory teams: "The reality is we're going to make advisor teams from pick-up teams in units [leftover remnants after other units have been formed]. It's all about mission accomplishment. If you want to have a good team, [you need quality people]. But hey, that's what we do." At another point in our interview, he added, "The Marine Corps tries to fill the need for advisors with augments [supplemental Marines who are not assigned permanently to a billet in an organization]. [But] we've got to get away from finding guys as additional augments."

Although this ad hoc approach to staffing these programs has its disadvantages, there are also advantages to bringing in augments, particularly from the reserves. As Master Gunnery Sergeant Rojas narrated in his interview above, reservists often have better people skills due to their experiences in their civilian lives at home. Major Neal also expressed this belief that reservists can bring more to culturally oriented programs, "We have none of the capabilities that the Army has for civil affairs and stability ops [operations].

But when you bring in reservists, you get people who are so much more accustomed to doing this. [For example, with Marines, maybe] you're doing a patrol in the city. . . . And a group drops down on them. The Marines will shoot them out. But if you have a CAG [civil affairs group], they might have worked it out otherwise."

Despite the special skills that reservists bring to the new MOSs and programs, or perhaps exactly because they bring unique skills—defining them as individuals rather than part of the team—these culturally oriented MOSs fall low in the Marine Corps cultural hierarchy that idolizes the Marine "on the ground" and particularly the infantry. As Lieutenant Colonel Terlizzi observes in a publication on cultural competence and civil-military operations,

> All services have hierarchies. In the Marine Corps it tends to be the combat arms specialties. While Marines of every occupation are taught that they are all rifleman, it has been the infantry that has held the place of prominence in our hierarchical order. As a result biases based on occupational specialties prevail. Take, for instance, CA (civil affairs) Marines and/or military information support operations (MISO) Marines. . . . These specialties are inherently focused on cross-culture considerations, yet as we decrease our combat participation in Afghanistan there is debate as to their utility in the future.[12]

Perhaps Colonel Leeds, who had served on the G-9 (joint civil affairs) staff in Afghanistan, described it most colorfully: "The 3 [infantry] are the meat eaters and the 9 [civil affairs] are the vegetarians. That's how the Marine Corps thinks." To understand these comments, one must of course realize that in the Marine Corps, "real men eat meat."

Creating Regional Specialists to Provide a General Skill: The RCLF Program

Creating supplementary MOSs and programs with a people-centered focus has been one approach to provide needed cultural skills to the Corps, without requiring a major shift in the way the Corps conducts business. RCLF—the Corps' Regional, Culture, and Language Familiarization program--has been the second solution. In contrast to the MOS/program approach, which has fostered the development of a "culture general" set of skills enabling Marines to work with the people in any area, RCLF focuses on creating a cadre of Marines with specific regional and language skills. Paradoxically, however, the design of the program is not intended to make regional or linguistic specialists, but rather to create a general pool of Marines who can be of use wherever they deploy.

The RCLF concept is simple and fits within existing Marine Corps training, education, and promotion tracks. More important, it does not interfere in any way with M&RA's system of personnel assignments. Instead, RCLF *reshapes* cultural specialists to fit within the Marine ethos of "every Marine a rifleman."

As 26th MEU's deployment (see chapter 3) illustrates, it is neither practical nor realistic to believe that Marines on that MEU could have possibly developed all the specific language and cultural skills necessary for a nine-month float that included operations in Pakistan, Afghanistan, Lebanon, and Libya as well as military training exercises with three additional Middle Eastern and African countries. Yet given DoD and Marine Corps policies requiring the Corps to create a culturally and linguistically capable force, a solution had to be found. The result was quite ingenious.

The Marine Corps document describing RCLF states:

> This initiative, the Regional, Culture and Language Familiarization (RCLF) Program, is designed to provide the initial foundation for a cross-culturally competent GPF (general purpose force) with diverse regional understanding and functional language capacity. . . . It is not intended to produce LRC (language, region and culture) expertise and is separate and distinct from other programs designed to produce a small cadre of professionals such as our international officers, cryptolinguists, intelligence workforce, and others.[13]

The program assigns all career Marines (all officers and reenlisting Marines) to study the culture and language of one of seventeen different world regions. This program begins at The Basic School for officers and upon reenlistment for the enlisted Marines, continuing throughout the Marine's career. The assumption, however, is not that any *specific* Marine will ever be deployed to the region he or she has studied. In fact, the expectation is that a large percentage of the Marines will never deploy to the region they have studied. However, *statistically*, the odds are that for any given deployment at least *some* of the Marines in the deploying unit will know something about that region (one in seventeen, in fact). Thus when a unit is suddenly deployed to West Africa, or the Philippines, or perhaps to six or seven different locations on a MEU, it can still be culturally and linguistically prepared for each destination, for within each unit, according to statistical odds, there will always be some Marines who have studied that area and can provide assistance to the commander and their fellow Marines.

The result is a new cultural version of the "everymarine"—a Lance Corporal Binotz with a specific language and regional specialty in every unit. As with all other Marine ways of doing things, the particular individual possessing the specific regional and cultural knowledge is not important. But his or her very presence on the team strengthens the entire unit and increases its ability to handle each new cultural challenge.

One decade after the United States invaded Iraq, and hundreds of DoD policy documents on language and cultural specialists later, we find a fascinating paradox. The Marine Corps has indeed responded to these external directives, but not by changing its fundamental culture. Instead of adapting Corps structure and manpower assignments to develop, identify, and deploy Marines with specialized language and culture skills, the Corps has done the reverse. It has reshaped the new external demands for unique specialists: creating a Corps that has a body of "green" Marine riflemen that jointly, as a team, can provide the cultural expertise needed for any deployment to any location. Thus, ironically, the Marine Corps has developed a language and culture general purpose force that still has "no 'I' in team."

Conclusion

> There needs to be institutional change in the Marine Corps regarding how
> to build the capability for "soft skills." "Soft skills" should be better defined
> and woven into our doctrine. We still need to maintain a lethal and ruthless
> force, but we and our nation can no longer afford to break and kill everyone
> who poses a threat to us. We must have the capacity to empathize and
> understand our foes and others caught in the middle of troubled regions
> and help reconcile those who are not really our enemy. We need to have
> a better sense of history and maintain better continuity of the issues and
> relationships. [And] our personnel system must be modified to embrace
> those specialists that have the "soft skills" to use in planning, training/
> teaching and advising, and not penalize them.
> —*Marine Major and Military Police Officer (CAOCL survey)*

HOW DOES A MILITARY SERVICE FOUNDED ON THE principles of "locate, close with, and destroy the enemy" suddenly shift its identity to become one that also resolves conflict through "soft skills" such as negotiation, perspective taking, and cross-cultural understanding? As the major above points out, such a shift requires much more than adding a few Arabic or Pashtu language classes to the predeployment training schedule. Such a radical cultural adaptation in the way the Marine Corps fights wars requires fundamental institutional change—from developing new doctrine to reorienting training and education programs to a restructuring of manpower assignments and promotion standards. But such changes are not easy and certainly do not follow a predictable linear pattern.

This book explores the process of military cultural change from the inside—from the perspectives of individual Marines and programs in the Corps—rather than from the point of view of the external policy-making organizations. The previous chapters have examined what happens when a well-intended policy designed in Washington to improve the way the United

States fights in a new irregular warfare environment "hits the ground" and must be implemented by Marines in Iraq or Afghanistan or even Quantico, Virginia.

Do military organizations (and perhaps, for that matter, any organization or group) simply accept the new external change as mandated, shift around a few personnel members and organizational structures, write a few new directives, and voilà!—the result pops out of the military black box exactly (or very closely) as directed? And if not, is the failure due to unclear policy design, poor or ineffective leadership and communication, or the lack of sufficient stakeholders in the policy-making process? In fact, if the result is not exactly what is intended, but the solution seems to work for those in the organization, how do we evaluate the outcome? Has the policy failed or succeeded?

In this book, I argue that military and organizational change is not a simple unidirectional process—where changes in the external strategic culture of the state or society are merely mirrored in parallel shifts in the military (the prevailing view in the contemporary literature on military change today). As this study illustrates, military organizations are not inert black boxes composed of random collections of people from the same country who view and interact with the world in a similar manner to the civilian population. Rather, military change is an interactive process, in which external shifts and pressures from the state, society, and the battlefield are integrated and reworked into the unique internal cultural and structural patterns of the specific military organization. The result is a synthesis or mixing of external directives and realities with the existing internal organizational beliefs, identity, and ways of conducting business—a process that the Marines call "Marinization."

According to the words and stories of the Marines in the preceding pages, the Marine Corps is in many ways a distinct U.S. subculture and not merely a random collection of Americans holding military jobs. Marines maintain a clearly unique identity based on a proud history, a tightly integrated community and way of life, and their role as an expeditionary "emergency response" team. Marines speak their own language—"Marine speak"; they wear unique clothing and cut their hair in distinct ways; and they have specific Marine holidays and traditions such as the Birthday Ball. Indeed, if the Marine Corps were not a military organization, social scientists long ago would have categorized it as an ethnic group due to its unique language, dress, customs, history, and identity.

More significant than their outward signs of a distinct subculture, however, Marines in this study described a set of ideals and values that—while not incompatible with American beliefs—are certainly different from those of the average American civilian. In this book, I have used Marine language and sayings to provide a window into understanding Marine Corps values and identity. For example, Marine philosophies of "every Marine a rifleman" and "there's no 'I' in team" reflect a particularly egalitarian belief that the team is far more important than the individual. In contrast to generally American individualistic ideals, Marines emphasized the values of self-sacrifice and the ability to withstand hardship and pain; they viewed selfishness, greed, laziness, and self-aggrandizement as unacceptable flaws; and they worked valiantly to uphold the expectation that leaders demonstrate humbleness and concern for their troops, rather than using the position for personal gain.

On the other hand, the Marines I interviewed and observed took great pride in their role as members of an elite Corps. "Honor, courage, and commitment" was viewed as a foundational statement of core values that underlies their commitment to uphold and continue the great deeds and historical legacy of all Marines who have preceded them. Indeed, Marine stories and narratives reflected the belief that no one in the Corps every really dies or retires. "Once a Marine, always a Marine" was not just a saying but a reflection of the belief that all Marines form part of the body of the Corps (a rather interesting religious metaphor). In public speeches and eulogies, death, particularly in battle, was described not as the end of a Marine's life but rather as a continuation of the great and honorable bloodline of all Marines who have upheld the values of the Corps since 1775.

Finally, the Marines in this study described themselves as at the "tip of the spear" (at the forefront) of any situation. Due to their expeditionary nature as "soldiers of the sea," Marines explained, they must be poised to respond and adapt to any situation at a moment's notice. The Marine philosophy of "semper gumby" epitomizes this flexible identity, which is demonstrated not only in Marines' pride in their ingenuity but also in their flexible organizational structures based on the Marine Air Ground Task Force (MAGTF). Concerned for survival—the Corps has repeatedly been under budgetary threats from Congress—the Marines I interviewed described the importance of accomplishing the mission quickly and frugally in order to demonstrate their value to the American public. This cultural and organizational emphasis leads to a very pragmatic focus on getting things done ("where's the 'so what'?"); a Spartan "can-do" attitude to make do with whatever resources are available; and a willingness to take risks: focusing on "the 80 percent solution" rather than perfection.

Thus while the majority of research on the military presumes that military members hold the same values and view the world in the same way as civilians (because they all come from the same society), this study challenges this basic assumption. Indeed, according to my observations of Marines and civilian subject matter experts (SMEs) working together, one of the greatest challenges in incorporating cultural considerations in military operations was that civilians and Marines held very different assumptions about what culture was and why it was important in the battlespace. The result has been a set of sometimes hilarious, sometimes extraordinarily frustrating cross-cultural miscommunications and misunderstandings within the *same* American culture.

What happens when the Marines' distinct worldview, unusually flexible organizational structures, and unique set of values come into contact with a culture policy developed in Washington, D.C., by a set of civilians and military leaders who see the world in quite different ways? In the specific case of the Marine Corps analyzed here, I discovered a fascinating process of reshaping and integration between the two cultures of Washington, D.C., and the Marine Corps, resulting in a set of solutions that were neither specifically Marine nor civilian, but a creative synthesis of both worlds. This process, I argue, consisted of four distinct though overlapping patterns of adaptation: *simplification*, *translation*, *processing*, and *reshaping*.

As I describe in chapter 6, in the early days of combat in Iraq, Marines' initial response was to *simplify* the problem of cross-cultural interaction and to view it as just another familiar (though context specific) conflict situation to which Marines could quickly and easily adapt. Culture thus became defined as a short-term problem limited to the latest conflict rather than requiring a fundamentally different and new approach to resolving conflict. Correspondingly, challenges with interacting with the population were seen as the result of two familiar and common parallel problems—ethical and linguistic—both of which could be solved by better predeployment training.

As a result, early solutions simplified the problem and focused on "cultural sensitivity training": teaching Marines the proper "do's and don'ts" of acting appropriately in a Muslim country. Reflecting the Marine Corps' ideals of "semper gumby" and their role at the "tip of the spear" as leaders and innovators, individual Marines and their units developed rapid, ad hoc, creative short-term programs, turning to an enormous range of "experts"—from expatriate Iraqis to Arabic speakers from the Middle East and North Africa to university professors studying the region—to provide culture and language training as well as serving as interpreters and translators. The inherently flexible nature of Marine Corps organizational structures fostered the creation of a dizzying array of culture and language programs and projects. Although these rapid efforts did lead to some initial successes in theater, the result was an enormously confusing hodgepodge of translators, instructors, and programs using vastly different methods and approaches—a legacy that was later to hinder the Corps as it began to realize that the problem of culture was not going to simply disappear after Iraq.

Not surprisingly, after this initial surge of effort, many Marines became disillusioned with the results of the often extremely basic and inconsistent cultural and language training programs. Equally frustrating was the growing realization that the solutions to the problems in theater required more than simply learning a few words and phrases or telling Marines not to stare at veiled Muslim women. Recognizing the need to understand the more complex cultural dynamics of the situation, Marines turned to social scientists and other "cultural" experts to translate and explain what was happening with the population. This approach defined the problem as one of finding the correct *translation* of Iraqi (and later Afghan) culture into familiar American—and more specifically—Marine terms. Marines hoped that by finding the right experts who could translate the local "cultural code," they would be able to quickly find the "center of gravity" for the situation and develop the appropriate and effective solution.

Thus civilian SMEs were hired to develop computerized cultural and language modeling, analysis, and simulation programs; to work alongside Marines as advisors in theater; and to serve as part of the military planning teams. The expectation was that these SMEs, with their presumed bicultural skills, expertise, and insights, could provide a clear interpretation of events and guide Marines in making better operational decisions. In order to better understand the cultural and human dynamics in conflict environments, the Marine Corps even developed a new approach to planning, adding a "Green Cell" to staffs that focused on the needs of the population. Shifting the definition of culture from a problem of "do's and don'ts" (a tactical perspective) to one of understanding the larger social and cultural context (an operational and strategic problem requiring better planning and analysis) did result in a number of apparent successes in theater—most notably the alliance with tribal leaders to produce the "Anbar Awakening" in Iraq.

Ironically, however, while the Marines hired these SMEs to translate and interpret the foreign cultural situation for them, civilians and Marines themselves had differing cultural perspectives on their roles and the purpose of their work together. Focusing on Marine Corps ideals of quick and simple mission accomplishment, Marines sought a clear set of concrete solutions that could be applied on the ground immediately, while SMEs who valued careful scholarship were often primarily concerned with producing accurate, carefully analyzed products. As a result, frequently the cooperation between civilians and Marines led to confusion and misunderstandings—and the culture of the population became lost in translation.

Simplification and translation can both be seen as short-term processes of adaptation—quick and easy efforts to resolve the current problem without requiring radical change within the Marine Corps itself. With time, however, as the Marine Corps shifted its combat efforts from Iraq to Afghanistan, it became evident that the cultural aspects of conflict continued to be an enduring issue. As a result, the Marines adopted yet a third, more long-term approach to addressing the cultural aspects of conflict: *processing*. Recognizing the limitations and problems resulting from the inconsistent ad hoc methods initially used to develop culture and language programs, the Marine Corps Center for Advanced Operational Culture Learning (CAOCL) developed a language and culture training process that was compatible with and followed accepted and standardized Marine Corps teaching and training methods. This approach embedded culture and language training "into the very fabric of the Corps" by processing the culture training to "look, smell, and taste Marine."

Given Marine Corps ideals of "every Marine a rifleman," CAOCL created a "one-size-fits-all" standardized training package that would give all Marines a basic and consistent set of operationally focused culture and language skills for any region of the world to which they would deploy. Similar to any efficient manufacturing process, the intention was to streamline and make all culture and language training programs conform to an identical format—using the Marine Corps method of the systems approach to training (SATs). This standardized process resulted in a predictable program of instruction that could be distributed across the Marine Corps to any unit in any location and taught by any "qualified" Marine (just as the Corps distributes standardized training packages on how to shoot a rifle). However, by standardizing culture training to fit Marine Corps training processes, the more nuanced, holistic, and conceptual aspects of culture became lost, subsumed under a pedagogical structure that was better suited to training marksmanship than developing critical thinking about foreign populations.

The fourth and most recent approach that the Marine Corps has taken to adapt to the external Washington, D.C. cultural policy imperatives has been *reshaping*. Like processing, reshaping reflects a longer-term institutional response to external pressures for change. Yet reshaping also allows the Corps to adapt to the external pressures for change without radically altering Marine Corps structures or ways of conducting business. Instead, reshaping molds and restructures the external initiative to fit into the culture and structure of the organization—changing the policy rather than the organization.

Reshaping is most evident in the Marine Corps' efforts to fit culture and language specialists (as required by DoD) into an egalitarian culture where specialists are seen as antithetical to building a cohesive flexible team whose members are "jacks of all trades." Although the Marine Corps does have a FAO (foreign area officer) and RAO (regional affairs officer) program as well as a clear cadre of heritage language–speaking Marines from around the world, it has not shifted manpower assignments to deploy individual Marines with specific culture and language skills to needed areas.

Given the Marine Corps' expeditionary nature—deploying anywhere around the world at a moment's notice—the Corps has argued that it would be impractical to spend years of time and money on training Marines to become fluent in the many hundreds of languages spoken around the world. So, although the U.S. Congress had mandated that all the U.S. military services focus their efforts on language training, the Marine Corps' response to this policy has been to develop a "culture general" training program that could be used by Marines to prepare for deployment to any country around the world.

The Marine Corps' emphasis on culture rather than language is a pragmatic adaptation to the Corps' operational realities. By focusing on developing general cross-cultural skills that can be applied to working with all people around the world regardless of their specific language or customs, the Marine Corps has sought to meet the mission (work effectively with foreign populations) while retaining its nonspecialist "semper gumby" culture and way of operating.

Accordingly, the Corps has developed what would seem to be a paradoxically illogical but rather creative program called the Regional, Culture, and

Language Familiarization (RCLF) program to address the external policy requirements. RCLF relies on the statistical probability of a group rather than the special skills of an individual to assure that the Marine Corps will always have some Marines with the appropriate cultural and language expertise in any region. The program works by assigning every career Marine to study one of seventeen regions (and associated languages of the world). The assumption is that in any deploying unit, regardless of where the unit goes, one in seventeen Marines will have had some basic culture and language training on the region, which he or she can then impart to his or her fellow Marines. In this solution, culture specialists become reshaped into Marine Corps ways of conducting business, rather than the reverse: preserving Marine Corps culture, while responding at least to the perceived intent of the external congressional and Department of Defense policy imperatives.

By examining military change from the "emic" perspective of Marines, rather than from an "etic," or external, perspective, this study challenges and refines a number of commonly accepted theories about military change and policy implementation. First, although there certainly is sufficient evidence to indicate that changes in the surrounding society and state or strategic culture are reflected in changes in military culture, this research suggests that these changes are by no means linear. As the case of the Marine Corps indicates, scholars must seriously reassess the untested assumption that members of the military are simply mirrors of society who respond to social and political changes in similar and predictable ways to their civilian counterparts.

Although to date there are only a bare handful of ethnographic "insider" studies of other military services, the data from these studies indicate that the Marine Corps is not the only military to have a distinct culture. In fact, the research by Simons,[1] Fujimura,[2] Tortorello,[3] and Ben Ari[4] overwhelmingly suggests that members of other militaries also hold unique identities, values, and worldviews that differ significantly from the surrounding civilian society. Given the obvious need for more detailed ethnographic research on internal military cultures, it is my hope that this book will encourage some members of the anthropological community to reconsider the prevailing animosity to conducting intensive ethnographic fieldwork among and within military communities. My research also challenges a second assumption in most studies of military change: that militaries are undefined "black boxes" whose activities are dictated and controlled by senior leadership and external policy makers. Again, although there is ample evidence in the political science and history literatures that famous military and state leaders can and do create dramatic changes in military structures and cultures, this research suggests that change is not unidirectional. Although military adaptation can be forced from the top down, this study illustrates that change also originates from the bottom up, independent of and even contrary to senior and external guidance. Russell's[5] study of change and adaptation among U.S. military units in Iraq provides similar evidence for bottom-up adaptation and change within the Army and the Marine Corps. These findings on military organizational change are congruent with the current literature on organizational change in civilian groups suggesting that organizations (whether military or civilian) are complex adaptive structures consisting of individuals and subcultures that often resist, redefine, and respond to external mandates in creative and innovative ways.

Third, this study indicates that policy scholars need to include research on the internal dynamics of organizations *receiving* the policy. As the case of the Marine Corps demonstrates, recipient organizations can play active roles in interpreting and applying external policies, adapting the external directives to fit within their own cultures. Indeed, as Yanow observes in her book *How Does a Policy Mean?*,[6] policy makers need to consider how organizational culture influences the way that policies are understood and translated into action.

Although this is a case study of one military organization, the patterns of adaptation that I observed in the Marine Corps may be common across other U.S. military services, and even across organizations in general. Although I can only offer anecdotal evidence, my own experience working with the other U.S. military services over the past seven years indicates that several of the four patterns I identified were also employed by the Army, Navy, and Air Force in response to the same DoD culture policies. Like the Marine Corps, the Army approached early operations in Iraq by *simplifying* the problem and then, later, seeking to *translate* culture by the use of SMEs. And early on, both the Air Force and the Navy began programs to *process* culture and language into familiar formats. In the case of the Navy, culture cards and classes were developed that clearly resembled the familiar disembarkation lectures that sailors received regarding appropriate behavior in port. And the Air Force focused on embedding its culture and language programs into the professional military education curricula, tying accreditation to their implementation, and ultimately linking culture and language education to foundational

Air Force documents and processes. Perhaps because I am not intimately familiar with the cultures of the sister military services, I do not know of any clear cases of *reshaping* in the other services. These anecdotal observations leave much room for future research on U.S. military organizational cultures, little of which is understood to date.

Finally, and perhaps most important, this study forces scholars and policy experts alike to rethink their definitions of successful policy implementation. As the Marine Corps case illustrates, what may seem to policy makers and researchers to be a failure or a misapplication of policy may be interpreted by the recipient organization as an effective and appropriate solution to the requirement. In fact, it is likely that one of the greatest problems with policy implementation and evaluation is not that it is implemented "incorrectly," but that we have failed to understand that there are competing cultural perspectives on what "success" means.

Thus, in the case of the Marine Corps, the seemingly paradoxical "Marinization" of cultural skills—creating a generalized pool of basically culturally competent Marines instead of building a Corps made of increasingly specialized regional and language experts—runs in the face of our competitive, individualistic American culture. In a civilian culture that rewards outstanding unique Americans, idolizes stars, and demands increasing specialization and expertise to succeed in academia and business, the Marine Corps' focus on building the team to the detriment of the expert specialist seems illogical and/or inefficient.

But such a judgment is based on an "ethnocentric" American cultural worldview rather than examining the world from the perspective of the "native culture"—that of the Corps. Ironically, and perhaps fittingly, the true cultural conflicts may not be between the Marines and the "strange" peoples in Iraq, Afghanistan, and other foreign lands to which Marines deploy. The real culture battles are being fought here on American soil—between an American academic, government, and defense culture that values and expects solutions to its conflicts to be provided through increasingly narrow specialization and technical expertise and a Marine Corps whose success has been based on another set of ideals: teamwork, flexibility, self-sacrifice, a focus on the humble guy on the ground, and, most important, accomplishing the mission, regardless of how unconventional the methods may be to get there.

Appendix: Marine Corps Ranks and Organizational Structures

TABLE A1 Marine Corps ranks and insignia

Enlisted

N/A	PVT	Private	E-1
	PFC	Private First Class	E-2
	LCPL	Lance Corporal	E-3
	CPL	Corporal	E-4
	SGT	Sergeant	E-5
	SSGT	Staff Sergeant	E-6
	GYSGT	Gunnery Sergeant	E-7
	MSGT	Master Sergeant	E-8
	1ST SGT	First Sergeant	E-8
	MGYSGT	Master Gunnery Sergeant	E-9
	SGTMAJ	Sergeant Major	E-9
	SMMC	Sergeant Major of the Marine Corps	E-9

TABLE A1 Marine Corps ranks and insignia (continued)

Officer

(GOLD)	2NDLT	Second Lieutenant	O-1
(SILVER)	1STLT	First Lieutenant	O-2
(SILVER)	CAPT	Captain	O-3
(GOLD)	MAJ	Major	O-4
(SILVER)	LTCOL	Lieutenant Colonel	O-5
	COL	Colonel	O-6
	BGEN	Brigadier General	O-7
	MAJGEN	Major General	O-8
	LTGEN	Lieutenant General	O-9
	GEN	General	O-10

Warrant Officer

	WO	Warrant Officer	W-1
	CWO2	Chief Warrant Officer 2	W-2
	CWO3	Chief Warrant Officer 3	W-3
	CWO4	Chief Warrant Officer 4	W-4
	CWO5	Chief Warrant Officer 5	W-5

SOURCE: Adapted from: U.S. Marine Corps, "Marine Corps Ranks," *Marines: The Official Site of the Marine Corps*

TABLE A2 Marine Corps organizational structures

	(Ideal/Theoretical Configurations of MAGTFs)		
	Marine Expeditionary Unit (MEU)	*Marine Expeditionary Brigade (MEB)*	*Marine Expeditionary Force (MEF)*
Command	Colonel	Brigadier General	Lieutenant General
Ground Combat Element (GCE)	Battalion Landing Team (BLT)	Regimental Landing/ Combat Team (RLT/ RCT)	Marine Division (Division)
Aviation Combat Element (ACE)	Compsite Squadron (ACE)	Marine Aircraft Group (MAG)	Marine Aircraft Wing (MAW)
Logistics Combat Element (LCE)	Combat Logistics Battalion (CLB)	Combat Logistics Regiment (CLR)	Marine Logistics Group (MLG)
Personnel	2,200	15,000–17,000	50,000
Amphibious Ships	3	17	30+

SOURCE: Adapted from "A Basic Guide to Working with the United States Marine Corps," 11; and U.S. Marine Corps, "Concepts and Programs," Headquarters Marine Corps, 2011.

Notes

Prologue

1. Names of all individuals, groups, and towns have been changed to protect the identity of the participants.

Introduction

1. Then Lieutenant General Mattis.

2. This saying is a foundational part of the Marine Corps infantry mission statement. See, for example, U.S. Marine Corps website, "Roles in the Corps: Infantry," www.marines.com/being-a-marine/roles-in-the-corps/ground-combat-element/infantry.

3. For a nice discussion of the two approaches, see Michael Hill and Peter Hupe, *Implementing Public Policy: An Introduction to the Study of Operational Governance*, 2nd ed. (Thousand Oaks, Calif.: Sage, 2008).

4. Originating from linguistic studies in the 1950s, the emic and etic distinction (deriving from phonemic and phonetic) emphasizes a methodological and epistemological contrast between native/internal and outsider/external research perspectives. For a discussion of the difference between emic versus etic approaches, see James P. Spradley, *Participant Observation* (Belmont, Calif.: Wadsworth, 1980).

5. Clifford Geertz, *The Interpretation of Cultures* (New York: Basic Books, 1977).

6. Joseph L. Soeters, "Culture in Uniformed Organizations," in *Handbook of Organizational Culture and Climate*, ed. Neal M. Ashkanasy, Celeste Wilderom, and Mark F. Peterson (Thousand Oaks, Calif.: Sage, 2000), refers to militaries and other uniformed organizations as an intermediate level between organizational culture and

national culture—since the environments of these organizations reach beyond the workday aspects of life into home, social, and other nonwork dimensions of daily life.

7. Lila Abu Lughod, *Veiled Sentiments: Honor and Poetry in a Bedouin Society* (Berkeley: University of California Press, 1986).

Chapter 1

1. See Paula Holmes-Eber, "A Day in the Life of the Marine Corps Professor of Operational Culture," in *Practicing Military Anthropology: Beyond Expectations and Traditional Boundaries*, ed. Robert Rubinstein, Kerry Fosher, and Clementine Fujimura (Sterling, Va.: Kumarian Press, 2013), for a description of some of these challenges.

2. John A. Nagl, *Learning to Eat Soup with a Knife: Counterinsurgency Lessons from Malaya and Vietnam* (Chicago: University of Chicago Press, 2005).

3. See, for example, Theo Farrell and Terry Terriff, eds., *The Sources of Military Change: Culture, Politics, Technology* (Boulder, Colo.: Lynne Rienner, 2001); Dima Adamsky, *The Culture of Military Innovation: The Impact of Cultural Factors on the Revolution in Military Affairs in Russia, the US, and Israel* (Stanford, Calif.: Stanford Security Studies, 2010); Keith Shimko, *The Iraq Wars and America's Military Revolution* (Cambridge: Cambridge University Press, 2010); Benjamin Buley, *The New American Way of War: Military Culture and the Political Utility of Force* (London: Routledge, 2008); James A. Russell, *Innovation, Transformation, and War: Counterinsurgency Operations in Anbar and Ninewa Provinces, Iraq, 2005–2007* (Stanford, Calif.: Stanford Security Studies, 2010); Eliot A. Cohen, "Change and Transformation in Military Affairs," *Journal of Strategic Studies* 27, no. 3 (2004); Robert M. Cassidy, *Counterinsurgency and the Global War on Terror: Military Culture and Irregular War* (Stanford, Calif.: Stanford Security Studies, 2008);Thomas Mahnken, *Technology and the American Way of War since 1945* (New York: Columbia University Press, 2010); Stephen P. Rosen, ed., *Winning the Next War: Innovation and the Modern Military*, Cornell Studies in Security Affairs (Ithaca, N.Y.: Cornell University Press, 1994); Jeremy Black, *War and the Cultural Turn* (Cambridge: Polity, 2010);Theo Farrell, Terry Terriff, and Osinga Frans, eds., *A Transformation Gap: American Innovations and European Military Change* (Stanford, Calif.: Stanford Security Studies, 2010).

4. For a detailed comparison of different ways of thinking about and defining strategic cultures, see Lawrence Sondhaus, *Strategic Studies and Ways of War* (London: Routledge, 2006).

5. Adamsky, *Culture of Military Innovation*.

6. Cassidy, *Counterinsurgency and the Global War on Terror*.

7. Farrell and Terriff, *Sources of Military Change*.

8. See, for example, Thomas A. Birkland, *An Introduction to the Policy Process: Theories, Concepts and Models of Public Policy Making*, 3rd ed. (Armonk, N.Y.: M. E. Sharpe, 2011); William N. Dunn, *Public Policy Analysis*, 5th ed. (Boston: Pearson, 2012); Kevin B. Smith and Christopher W. Larimer, *The Public Policy Theory Primer*

(Boulder, Colo.: Westview Press, 2009); John Kingdon, *Agendas, Alternatives and Public Policies*, 2nd ed. (New York: Longman, 2003); Michael E. Kraft, *Public Policy: Politics, Analysis and Alternatives*, 3rd ed. (Washington, D.C.: CQ Press, 2009); Paul Sabatier, ed., *Theories of the Policy Process* (Boulder, Colo.: Westview Press, 2007).

9. Sam C. Sarkesian, John Allen Williams, and Stephen J. Cimbala, *U.S. National Security: Policymakers, Processes and Politics*, 4th ed. (Boulder, Colo.: Lynne Rienner, 2008).

10. Hill and Hupe, *Implementing Public Policy*.

11. For overviews of anthropological approaches to policy studies, see Janine R. Wedel et al., "Toward an Anthropology of Public Policy," *Annals of the American Academy of Political and Social Science* 600 (2005); Chris Shore, Susan Wright, and David Pero, eds., *Policy Worlds: Anthropology and the Analysis of Contemporary Power* (Oxford: Berghahn Books, 2011); Anne Francis Okongwu and Joan P. Mencher, "The Anthropology of Public Policy: Shifting Terrains," *Annual Review of Anthropology* 29 (2000).

12. Jeffrey L. Pressman and Aaron Wildavsky, *Implementation: How Great Expectations in Washington Are Dashed in Oakland; or Why It's Amazing That Federal Programs Work at All* (Berkeley: University of California Press, 1984).

13. See, for example, Hill and Hupe, *Implementing Public Policy*.

14. Michael Lipsky, *Street Level Bureaucracy: Dilemmas of the Individual in Public Service*, 30th anniversary edition ed. (New York: Russell Sage Foundation, 2010).

15. Dvorah Yanow, *How Does a Policy Mean? Interpreting Policy and Organizational Actions* (Washington, D.C.: Georgetown University Press, 1997).

16. Kate Ingram, "Equal Opportunities, the Clothing Industry and the Law," in *Culture and Policy in Northern Ireland: Anthropology in the Public Arena*, ed. Hastings Donnan and Graham Farlane (Chester Springs, Pa.: Dufour Editions, 1997).

17. For an introduction to the various aspects of these approaches, see, for example, Thomas W. Britt, Amy B. Adler, and Carl A. Castro, eds., *Military Life: The Psychology of Serving in Peace and Combat*, vol. 4 (Westport, Conn.: Praeger Security International, 2006); Janice H. Laurence and Michael D. Matthews, eds., *The Oxford Handbook of Military Psychology* (Oxford: Oxford University Press, 2011); Guiseppe Caforio, ed., *Handbook of the Sociology of the Military* (New York: Springer, 2006).

18. See, for example, Aaron Belkin, *Bring Me Men: Military Masculinity and the Benign Facade of American Empire, 1898–2001* (New York: Columbia University Press, 2012); Melissa T. Brown, *Enlisting Masculinity: The Construction of Gender in US Military Recruiting Advertising during the All-Volunteer Force*, Oxford Studies in Gender and International Relations (Oxford: Oxford University Press, 2012); Penny F. Pierce, "The Role of Women in the Military," in Britt, Adler, and Castro, *Military Life*.

19. See, for example, Karin De Angelis and David R. Segal, "Minorities in the Military," in Laurence and Matthews, *Oxford Handbook of Military Psychology*.

20. See, for example, Armando X. Estrada, "Gay Service Personnel in the U.S. Military: History, Progress and a Way Forward," in Laurence and Matthews, *Oxford Handbook of Military Psychology*; Gregory M. Herek and Aaron Belkin, "Sexual

Orientation and Military Service: Prospects for Organizational and Institutional Change in the United States," in Britt, Adler, and Castro, *Military Life.*

21. See James G. Hunt and Robert L. Phillips, "Leadership in Battle and Garrison: A Framework for Understanding the Differences and Preparing for Both," in *Handbook of Military Psychology*, ed. Reuven Gal and A. David Mangelsdorff (Chichester: John Wiley and Sons, 1991); Marina Nuciari, "Models and Explanations for Military Organization: An Updated Reconsideration," in Caforio, *Handbook of the Sociology of the Military*; and Joseph L. Soeters, Donna J. Winslow, and Alise Weibull, "Military Culture," in Caforio, *Handbook of the Sociology of the Military*, for discussions of the differences between "hot" and "cold" organizations.

22. Soeters, Winslow, and Weibull, "Military Culture."

23. Joseph L. Soeters and R. Recht, "Culture and Discipline in Military Academies: An International Comparison," *Journal of Political and Military Sociology* 26, no. 2 (1998).

24. Stasiu Labuc, "Cultural and Societal Factors in Military Organizations," in Gal and Mangelsdorff, *Handbook of Military Psychology.*

25. Joseph L. Soeters, Cristina-Rodica Poponete, and Joseph T. Page Jr., "Culture's Consequences in the Military," in Britt, Adler, and Castro, *Military Life.*

26. Charlie Moskos, "Institutional and Occupational Trends in Armed Forces: An Update," *Armed Forces and Society* 12, no. 3 (1986); Charlie Moskos and James Burk, "The Postmodern Military," in *The Adaptive Military: Armed Forces in a Turbulent World*, ed. James Burk (Boulder, Colo.: Westview Press, 1994); Charlie Moskos, John A. Williams, and David R. Segal, eds., *The Postmodern Military* (Oxford: Oxford University Press, 2000).

27. Charlie Moskos, "From Institution to Occupation: Trends in Military Organizations," *Armed Forces and Society* 4, no. 1 (1977): 41.

28. See George R. Lucas, ed., *Anthropologists in Arms: The Ethics of Military Anthropology* (Walnut Creek, Calif.: AltaMira Press, 2009), for a discussion of the ethical issues; and David H. Price, *Anthropological Intelligence: The Deployment and Neglect of American Anthropology in the Second World War* (Durham, N.C.: Duke University Press, 2008), for a discussion of the historical relationship between anthropology and the security communities.

29. Laura Nader, "Up the Anthropologist—Perspectives Gained from Studying Up," in *Reinventing Anthropology*, ed. Dell H. Hymes (New York: Pantheon Books, 1972).

30. See Hugh Gusterson, "Studying up Revisited," *PoLAR: Political and Legal Anthropology Review* 20, no. 1 (1997).

31. Robert Rubinstein, *Peacekeeping under Fire: Culture and Intervention* (Boulder, Colo.: Paradigm, 2008), 55.

32. See, for example, Network of Concerned Anthropologists, *The Counter-Counterinsurgency Manual* (Chicago: Prickly Paradigm Press, 2009); Rochelle Davis, "Culture as a Weapon System," *Middle East Report* 255 (Summer 2010); Dan Vergano and Elizabeth Weise, "Should Anthropologists Work Alongside Soldiers? Arguments Pro, Con Go to Core Values of Cultural Science," *USA Today*, December 9, 2008.

33. Robert Albro et al., eds. *Anthropologists in the Securityscape: Ethics, Practice and Professional Identity* (Walnut Creek, Calif.: Left Coast Press, 2012).

34. See, for example, Paula Holmes-Eber, "Teaching Culture at Marine Corps University," in Albro et al., *Anthropologists in the Securityscape*; Robert Rubinstein, "Master Narratives, Retrospective Attribution and Ritual Pollution in Anthropology's Military Engagements with the Military," in Rubinstein, Fosher, and Fujimura, *Practicing Military Anthropology*.

35. For a good introduction to the literature, see Anna Simons, "War: Back to the Future," *Annual Review of Anthropology* 28 (1999); and Hugh Gusterson, "Anthropology and Militarism," *Annual Review of Anthropology* 36 (2007). See also Athina Karatzogianni, ed., *Violence and War in Culture and the Media: Five Disciplinary Lenses* (London: Routledge, 2012); Christos Kryou and Robert Rubinstein, "Cultural Anthropology Studies of Conflict," in *Encyclopedia of Violence, Peace and Conflict*, ed. Lester Kurtz (Amsterdam: Elsevier, 2008); Keith F. Otterbein, *The Anthropology of War* (Long Grove, Ill.: Waveland Press: 2009); Koen Stroeken, ed., *War, Technology, Anthropology* (New York: Berghahn Books, 2012); R. Brian Ferguson and Neil L. Whitehead, eds., *War in the Tribal Zone: Expanding States and Indigenous Warfare* (Santa Fe, N.Mex.: School of American Research Press, 1992), for good introductions to the field.

36. See Catherine Lutz, *Homefront: A Military City and the American Twentieth Century* (Boston: Beacon Press, 2002); Catherine Lutz, ed., *The Bases of Empire: The Global Struggle against Military Posts* (New York: NYU Press, 2009); Pamela R. Frese, "Guardians of the Golden Age: Custodians of U.S. Military Culture," in *Anthropology and the United States Military: Coming of Age in the Twenty-First Century*, ed. Pamela R. Frese and Margaret C. Harrell (New York: Palgrave-Macmillan, 2003); Margaret C. Harrell, "Gender and Class-Based Role Expectations for Army Spouses," in Frese and Harrell, *Anthropology and the United States Military*; Antonius C. G. M. Robben, ed., *Iraq at a Distance: What Anthropologists Can Teach Us About the War* (Philadelphia: University of Pennsylvania Press, 2010).

37. Price, *Anthropological Intelligence*; Robert Albro, "Public Anthropology and Multitrack Dialoguing in the Securityscape," in Albro et al., *Anthropologists in the Securityscape*; Keith Brown, "'All They Understand Is Force': Debating Culture in Operation Iraqi Freedom," *American Anthropologist* 110, no. 4 (2008); Brown, *Enlisting Masculinity*.

38. Rubinstein, "Master Narratives."

39. Celso Castro, "Anthropological Methods and the Study of the Military: The Brazilian Experience," in *Qualitative Methods in Military Studies*, ed. Helena Carreiras and Celso Castro (London: Routledge, 2012); Piero C. Lierner, "Side Effects of the Chain of Command on Anthropological Research: The Brazilian Army," in Carreiras and Castro, *Qualitative Methods in Military Studies*; Alejandra Navarro, "Negotiating Access to an Argentinean Military Institution in Democratic Times: Difficulties and Challenges," in Carreiras and Castro, *Qualitative Methods in Military Studies*.

40. Charles Kirke, "Insider Anthropology: Theoretical and Empirical Issues for the Researcher," in Carreiras and Castro, *Qualitative Methods in Military Studies*.

41. Neta Bar and Eyal Ben-Ari, "Israeli Snipers in the Al-Aqsa Intifada: Killing, Humanity and Lived Experience," *Third World Quarterly* 26, no. 1 (2005); Eyal Ben-Ari, "Masks and Soldiering: The Israeli Army and the Palestinian Uprising," *Cultural Anthropology* 4, no. 4 (1989); Eyal Ben-Ari, *Mastering Soldiers: Conflict, Emotions and the Enemy in an Israeli Military Unit* (Oxford: Bergham Books, 1998); Eyal Ben-Ari and Sabine Fruhstuck, "The Celebration of Violence: A Live-Fire Demonstration Carried Out by Japan's Contemporary Military," *American Ethnologist* 30, no. 4 (2003).

42. Anna Simons, *The Company They Keep: Life Inside the U.S. Army Special Forces* (New York: Free Press, 1997).

43. Frank J. Tortorello, "An Ethnography of 'Courage' among U.S. Marines" (Ph.D. diss., University of Illinois at Urbana-Champaign, 2010).

44. Carol Burke, *Camp All American, Hanoi Jane and the High-and-Tight: Gender, Folklore and Changing Military Culture* (Boston: Beacon Press, 2005).

45. Hugh Gusterson, *Nuclear Rites: A Weapons Laboratory at the End of the Cold War* (Berkeley: University of California Press, 1998).

46. Clementine Fujimura, "Culture in/Culture of the United States Naval Academy," in Albro et al., *Anthropologists in the Securityscape*; Clementine Fujimura, "'Living the Dream': One Military Anthropologist's Initiation," in Rubinstein, Fosher, and Fujimura, *Practicing Military Anthropology*.

47. Kerry Fosher, *Making Homeland Security at the Local Level* (Chicago: University of Chicago Press, 2009).

48. For a comparison of these approaches, see Majken Schultz, *On Studying Organizational Cultures: Diagnosis and Understanding* (Berlin: Walter de Gruyter, 1995).

49. See Edgar H. Schein, *Organizational Culture and Leadership*, 3rd ed. (San Francisco: Jossey-Bass, 2004), for a good introduction to organizational studies of leadership.

50. Victor H. Krulak, *First to Fight: An Inside View of the Marine Corps* (Annapolis, Md.: U.S. Naval Institute Press, 1989).

51. Terry Terriff, "Warriors and Innovators: Military Change and Organizational Culture in the US Marine Corps," *Defence Studies* 6, no. 2 (2006).

52. Soeters, "Culture in Uniformed Organizations."

53. James G. Pierce, *Is the Organizational Culture of the U.S. Army Congruent with the Professional Development of Its Senior Level Officer Corps?* (Carlisle, Pa.: Strategic Studies Institute, U.S. Army War College, 2010).

54. Terry Terriff, "'Innovate or Die': Organizational Culture and the Origins of Maneuver Warfare in the United States Marine Corps," *Journal of Strategic Studies* 29, no. 3 (2006).

55. Frank Hoffman, "The Origins and Application of the Expeditionary Ethos" (Washington, D.C.: Scitor, 2007).

56. Ben Connable, "Culture Warriors: Marine Corps Organizational Culture and Adaptation to Cultural Terrain," *Small Wars Journal* (February 2008), http://smallwarsjournal.com/jrnl/art/culture-warriors.

57. James M. Smith, "Service Cultures, Joint Cultures and the US Military," *Airman-Scholar* (Winter 1998).

58. See Joanne Martin, *Cultures in Organizations: Three Perspectives* (Oxford: Oxford University Press, 1992); Joanne Martin, *Organizational Culture: Mapping the Terrain* (Thousand Oaks, Calif.: Sage, 2001).

59. Labuc, "Cultural and Societal Factors in Military Organizations."

60. Soeters, Winslow, and Weibull, "Military Culture."

61. Allan D. English, *Understanding Military Culture: A Canadian Perspective* (Montreal: McGill-Queen's University Press, 2004).

62. Mahnken, *Technology and the American Way of War.*

63. Rosen, *Winning the Next War.*

64. Mahnken, *Technology and the American Way of War*; Nagl, *Learning to Eat Soup with a Knife.*

65. For a particularly vicious example of the ostracism of anthropologists who work with the military, see Rubinstein, "Master Narratives."

66. For more in-depth discussions of the ethical aspects of working with the military, see Albro et al., *Anthropologists in the Securityscape*; and Holmes-Eber, "Teaching Culture at Marine Corps University." See also AAA Commission on the Engagement of Anthropology with the U.S. Security and Intelligence Communities (CEAUSSIC), "Final Report on the Army's Human Terrain System Proof of Concept Program," American Anthropological Association, October 14, 2009; Roberto J. Gonzalez, *American Counterinsurgency: Human Science and the Human Terrain* (Chicago: Prickly Paradigm Press, 2009); Lucas, *Anthropologists in Arms*; Network of Concerned Anthropologists, *Counter-Counterinsurgency Manual.*

67. In Holmes-Eber, "Teaching Culture at Marine Corps University," I describe some of my early, sometimes humorous, sometimes discouraging, challenges of communicating across the anthropological academic and military worlds.

68. I would like to thank Colonel (ret) Jeffrey Bearor, SES and executive director of Training and Education Command (TECOM); Major General (ret) Gardner, former president of MCU; Lieutenant General Murray, Commanding General TECOM; Colonel Jerry Wilson (ret), vice president MCU Academic Affairs; and Colonel (ret) George Dallas, director of CAOCL, for their encouragement, support, and mentorship of this project in its various stages over the years.

69. All research for this study was reviewed for human subjects protection by the Marine Corps Institutional Review Board (IRB).

70. Several of the interviews in the first eight months of this study were conducted in conjunction with Barak Salmoni, former deputy director of CAOCL.

71. It is important to note that due to ethical and professional considerations, I have never applied for or received a security clearance. Thus all data in this study, including interviews and discussions with military personnel, by definition, contain no secret information. With the exception of removing personally identifying information, then, no "scrubbing" of data from my interviews or fieldnotes has been necessary prior to their publication.

72. A separate IRB was obtained for this statistical study.

73. A large portion of the survey implementation, design, and analysis was conducted by Erika Tarzi and Basma Maki at CAOCL, and I am indebted to them for their diligent work. I am also deeply indebted to the hard work and support of the staff at Marine Corps Lessons Learned, who programmed and sent out the survey online for CAOCL. I would especially like to thank Ron Middlebrook for his extensive programming support.

74. Almost all Marines of higher ranks are included in GAL. However, entry-level Marines (E-1s privates, E-2s lance corporals, and O-1s second lieutenants) often do not obtain email addresses until they have completed their MOS training. General officers were excluded. A few Marine reservists who were currently on active duty did respond to the survey as well as seven Navy medical, engineering, and chaplain MOS who were embedded with Marine units.

75. For the purposes of this study, career Marines are defined as Marines who have completed their basic and MOS training and are actively involved in Marine Corps operations.

76. U.S. Marine Corps, "Concepts and Programs" (Headquarters Marine Corps, 2010).

77. U.S. Marine Corps Center for Advanced Operational Culture Learning (CAOCL), "2010 Survey of Marine Corps Attitudes to Culture and Language Training: Sample and Methods" (CAOCL, 2010).

78. Spradley, *Participant Observation*.

79. Paula Holmes-Eber, *Daughters of Tunis: Women, Family and Networks in a Muslim City* (Boulder, Colo.: Westview Press, 2003).

80. The extensive anthropological postmodernist literature on the limitations of colonialist anthropological research paradigms is far too long to cite here. However, a good start is Gaurav Gajanan Desai, *Postcolonialisms: An Anthology of Cultural Theory and Criticism* (New Brunswick, N.J.: Rutgers University Press, 2005). See also Fosher, *Making Homeland Security at the Local Level*; Gusterson, "Studying up Revisited"; and Rubinstein, *Peacekeeping under Fire*, for discussions of how shifting concepts of space and fieldwork and contemporary challenges in conducting anthropological fieldwork render traditional concepts of the participant-observer outdated.

81. Gusterson, "Studying up Revisited," 117.

82. For detailed discussions of the practical and ethical aspects of simultaneously practicing work and conducting anthropological research, see, for example, Gracia Clark, "Working the Field: Kumasi Central Market as Community, Employer and Home," *Anthropology of Work Review* 29, no. 3 (2008); Gusterson, "Studying up Revisited"; Rebecca Prentice, "Knowledge, Skill and the Inculcation of the Anthropologist: Reflections on Learning to Sew in the Field," *Anthropology of Work Review* 29, no. 3 (2008); Rubinstein, *Peacekeeping under Fire*; Gavin Whitelaw, "Learning from Small Change: Clerkship and the Labors of Convenience," *Anthropology of Work Review* 29, no. 3 (2008).

83. The survey data reported here belong to CAOCL and have formed part of a number of policy-oriented documents concerning the Marine Corps' progress in

implementing culture and language programs. However, the particular analyses of the survey data presented here, like my interpretation of the ethnographic study, are based on my own work.

Part I

1. Claude Levi-Strauss, *The Savage Mind*, trans. George Weidenfeld (Chicago: University of Chicago Press, 1962); Claude Levi-Strauss, *The Origin of Table Manners* (New York: Harper and Row, 1968).

2. Pierre Bourdieu, *Outline of a Theory of Practice*, trans. Richard Nice (Cambridge: Cambridge University Press, 1977); Pierre Bourdieu, *The Logic of Practice*, trans. Richard Nice (Stanford, Calif.: Stanford University Press, 1992).

3. William F. Hanks, "Pierre Bourdieu and the Practices of Language," *Annual Review of Anthropology* 34 (2005).

4. Noam Chomsky and Misou Rinat, *On Language: Chomsky's Classic Works Language and Responsibility and Reflections on Language in One Volume* (New York: New Press, 1998).

5. Naomi Quinn, "How to Reconstruct Schemas People Share from What They Say," in *Finding Culture in Talk: A Collection of Methods*, ed. Naomi Quinn (New York: Palgrave-Macmillan, 2005).

6. Martin, *Cultures in Organizations*; Martin, *Organizational Culture*.

Chapter 2

1. Howard L. Rosenberg, "Real Story of Jessica Lynch's Convoy," *ABC News Nightline* (June 17, 2003), http://abcnews.go.com/Nightline/story?id=128387&page=1.

2. Rosenberg, "Real Story of Jessica Lynch's Convoy," 2.

3. There is some debate as to the origin of this quote. While Marines typically reference General Gray when discussing the quote, the belief that all Marines should be riflemen existed long before General Gray's campaign. The slogan appears to have been in use long before the 1980s and popularized under General Gray.

4. U.S. Marine Corps Community Services, "The Marine Corps: A Young and Vigorous Force, Demographic Update" (2012).

5. U.S. Marine Corps Community Services, "Marine Corps," 8.

6. U.S. Marine Corps Community Services, "Marine Corps," 8.

7. "Intel" is short for the intelligence gathering military occupational specialty (MOS) and involves the task of researching the operational environment and understanding the adversary.

8. Note that the gendered use of "guy" is appropriate. Until recently, female Marines were not permitted in the infantry or other direct combat forces.

9. MOSs are all assigned a primary two-digit occupational field code.

10. In 2012, the Marine Corps began an experimental program to permit women to enter the Infantry Officers Course (IOC). For recent debates on the subject, see C. J. Chivers, "A Grueling Course for Training Marine Officers Will Open Its Doors to Women," *New York Times*, July 8, 2012; Katie Petronio, "Get Over It! We Are Not Created Equal," *Marine Corps Gazette* (2012), http://www.mca-marines.org/gazette/ article/get-over-it-we-are-not-all-created-equal.

11. We did in fact run that analysis, and in all statistical tests the ground combat forces used culture and language skills significantly more often than combat support and correspondingly valued these skills more than combat support.

12. According to the 2011 U.S. Marine Corps, "Concepts and Programs" (Headquarters Marine Corps, 2011), 36,966 Marines held the Primary Occupation Field of 03 (Infantry) out of a total force of 202,441.

13. In Marine Corps parlance, the word "vice" has come to replace the word "versus." Although the phrasing is grammatically incorrect, this book will use the language that reflects the way that Marines speak.

14. U.S. Marine Corps, "Facebook Page," http://www.facebook.com/#!/ marinecorps.

15. Max Walsh, "New USMC Recruiting Video," http://www.youtube.com/ watch?v=UFeHoMhuz7A.

16. According to the 2011 U.S. Marine Corps, "Concepts and Programs," 95,800 Marines, or 47 percent of the 202,441 total force, held one of the following two-digit occupational specialties: communications, artillery, engineer, tank, signals, ground electronics, motor transport, electronic maintenance, aircraft maintenance, avionics maintenance, airfield services, air control, navigation, and pilot.

17. According to the 2011 U.S Marine Corps, "Concepts and Programs," 35,195 Marines, or 17 percent of the 202,441 total force for that year, held aviation-related MOS.

18. The Marine Corps divides responsibilities for organizations along functional lines that are numbered.

Chapter 3

1. Santiago G. Colon, "Living on Ship: 26th MEU Marines Balance Duty and Downtime," http://www.26thmeu.marines.mil/News/NewsArticleDisplay/ tabid/2723/Article/65684/living-on-ship-26th-meu-marines-balance-duty-and-downtime.aspx.

2. "Marines Deliver Lifesaving Supplies to Flood Stricken Pakistanis," *Marine Corps Gazette* (January 9 2013), http://www.mca-marines.org/gazette/ marines-deliver-lifesaving-supplies-flood-stricken-pakistanis.

3. "26th MEU Slated to Deploy to Afghanistan," *Marine Corps Gazette* (2011), http://www.mca-marines.org/gazette/26th-meu-slated-deploy-afghanistan.

4. U.S. Marine Corps, "Lessons from 26th Marine Expeditionary Unit Operations," *Marine Corps Center for Lessons Learned 7*, no. 11 (November 2011).

5. See James F. Amos, "Role of the United States Marine Corps" (Memorandum for the Secretary of Defense from the Commandant of the Marine Corps, September 12, 2011); U.S. Marine Corps, "Vision and Strategy 2025" (Foreword by General James T. Conway, 2008).

6. U.S. Marine Corps, "Vision and Strategy 2025."

7. Pierce, *Is the Organizational Culture of the U.S. Army Congruent with the Professional Development of Its Senior Level Officer Corps?*; Schein, *Organizational Culture and Leadership*; Schultz, *On Studying Organizational Cultures*; Helen Schwartzman, *Ethnography in Organizations* (Newbury Park, Calif.: Sage, 1992).

8. Krulak, *First to Fight*, 1.

9. U.S. Congress, "National Security Act (Unification Bill)" (1947 [amended 1952]).

10. Adam Siegel, "A Chronology of U.S. Marine Corps Humanitarian Assistance and Peace Operations" (Alexandria, Va.: Center for Naval Analyses 1994).

11. U.S. Congress, "National Security Act (Unification Bill)."

12. U.S. Marine Corps, "Vision and Strategy 2025."

13. U.S. Marine Corps, "The Long War: Send in the Marines. A Marine Corps Operational Employment Concept to Meet an Uncertain Security Environment" (Quantico, Va.: U.S. Marine Corps, 2008).

14. U.S. Marine Corps Combat Training Command, "Marine Training Film."

15. Krulak, *First to Fight*.

16. In Amos, "Role of the United States Marine Corps," General Amos stated that the Marine Corps required only 7.8 percent of the total DoD budget.

17. The percentage of the U.S. military budget that the Marine Corps receives not only varies from year to year but also depends on whether specific types of funding, such as supplemental funds and OCO funds, are included.

18. Krulak, *First to Fight*.

19. Neil Bouhan and Paul Swartz, "Trends in U.S. Military Spending" (Maurice Greenberg Center for Geoeconomic Studies, Council on Foreign Relations, June 28, 2011).

20. Amos, "Role of the United States Marine Corps."

21. See, for example, http://patriot-nation.blogspot.com/2010/01/patriot-nation-motivational-posters.html.

22. Technically, fires include artillery, mortars, air firepower, and naval surface fires. Recently, cyber warfare has been added to the category.

23. Despite their amphibious identity, few Marines are actually trained in any form of naval skills (the main exceptions are those Marines trained to conduct amphibious landings using AMTRAKs—boats that can sail off from a ship and then drive on land when they reach the shore).

24. U.S. Marine Corps Force Structure Review Group, "Reshaping America's Expeditionary Force in Readiness: Report of the 2010 Marine Corps Force Structure Review Group" (Quantico, Va.: U.S. Marine Corps, March 14, 2011).

Chapter 4

1. Military time is calculated on a twenty-four-hour clock and reported as four digits—01–24 for the hour, followed by the minutes. So 0730 is 7:30 a.m.

2. In the Army and other services, the first phase of initial training of incoming recruits is called "boot camp"—and this has become the accepted term by the general public. However, the Marine Corps refers to initial training as "recruit training," which is conducted on "recruit depots."

3. U.S. Marine Corps Recruit Depot (MCRD), *Beginning the Transformation* (2008), 15.

4. Although entry into the Marine Corps is voluntary, a meaningful percentage of Marines come from Marine or military families—sometimes a hereditary pattern that can be continued for generations. It could be argued in this case that socialization into Marine Corps culture begins at birth.

5. This quote paraphrases the original famous quote by President Truman.

6. U.S. Marine Corps Recruit Depot (MCRD) Parris Island, "Command Brief" (2011).

7. U.S. Marine Corps Recruit Depot (MCRD) Parris Island, "Command Brief."

8. U.S. Marine Corps Recruit Depot (MCRD), *Beginning the Transformation*.

9. U.S. Marine Corps, "Your Impact," http://www.marines.com/global-impact/your-impact.

10. FallenUSSoldiers, "U.S. Marine Corps—Making a Marine," http://www.youtube.com/watch?v=MYRccSZgXV4; FallenUSSoldiers, "U.S. Marine Corps—Making a Marine Part 2," http://www.youtube.com/watch?v=aPgkACv7grk&feature=mfu_in_order&list=UL; FallenUSSoldiers, "U.S. Marine Corps—Making a Marine Part 3," http://www.youtube.com/watch?v=fPRzONdtgNo.

11. Marion F. Sturkey, *Warrior Culture of the U.S. Marines: Axioms for Warriors, Marine Quotations, Battle History, Reflections on Combat, Corps Legacy, Humor—and Much More—for the World's Warrior Elite* (Plum Branch, S.C.: Heritage Press International, 2003).

12. Actually, the correct German spelling would be Teufelshunde. However, the mistranslation has become part of Marine Corps culture, so I use the Americanization of the term here.

13. Sturkey, *Warrior Culture of the U.S. Marines*, 123.

14. OurMarines, "'No Compromises': Marine Commercial," http://www.youtube.com/watch?v=opAJtwyujvo.

15. marines, "Marine Corps Commercial: A Path for Warriors," http://www.youtube.com/watch?v=S_nApy9VVlk.

16. Marine OCS, "Marine Corps Commercial: Toward the Sounds of Chaos," http://www.youtube.com/watch?v=tYrBSTBHCS4.

17. U.S. Marine Corps, "Global Impact," http://www.marines.com/global-impact.

18. Walsh, "New USMC Recruiting Video."

19. U.S. Marine Corps Recruit Depot (MCRD), *Beginning the Transformation*, 15.

20. U.S. Marine Corps Recruit Depot (MCRD), *Beginning the Transformation*, 15.

21. U.S. Marine Corps Recruit Depot (MCRD) Parris Island, "Command Brief."

22. U.S. Marine Corps Recruit Depot (MCRD), *Beginning the Transformation*, v.

23. A safety glow belt that all Marines (and civilians) must wear when exercising on base so that cars can see them.

24. A long paved field used for drill practice, parades and shows, and graduation.

25. U.S. Marine Corps Recruit Depot (MCRD), "Modified 12 Week Recruit Training Schedule" (August 2011).

26. For a complete description of the Crucible, see James B. Woulfe, *Into the Crucible: Making Marines for the 21st Century* (Novato, Calif.: Presidio Press, 1998).

27. Although all of the other stories I have narrated were based on my own observations in the field, the Crucible was not being held during the week I came to visit at Parris Island. This story therefore is based on the YouTube video posted by FallenUS-Soldiers, "U.S. Marine Corps—Making a Marine Part 3."

Chapter 5

1. RSO literally stands for Range Safety Officer—someone who oversees range activities and safety.

2. Naval ranks are different than Marine Corps ranks. A naval captain is a very important person, equal to a Marine colonel. In the Marine Corps, a captain is a rather junior officer rank.

3. When the Marine Corps speaks of the strategic corporal, the Corps is referring to the impact that a corporal's actions could have at a national or even international level. A simple example is the international incident that occurred in 2012 due to the burning of Qurans in Afghanistan, conducted in large part by junior enlisted soldiers.

Part II

1. Jesse Singal, Christine Lim, and M. J. Stephey, "May 2003 Victory Lap," *Time*, March 19, 2010, http://www.time.com/time/specials/packages/article/0,28804,1967340_1967342_1967406,00.html.

2. "US Soldiers Killed in Baghdad," *BBC News* (July 7, 2003), http://news.bbc.co.uk/2/hi/middle_east/3050358.stm.

3. "Truck Bomb Kills Chief U.N. Envoy to Iraq: 17 Dead, 100 Injured by Explosion at U.N. Headquarters," *CNN* (August 20, 2003), http://www.cnn.com/2003/WORLD/meast/08/19/sprj.irq.main/.

4. "Red Cross Pulls Out of Baghdad, Basra," *ABC News* (November 8, 2003), http://www.abc.net.au/news/2003–11–08/red-cross-pulls-out-of-baghdad-basra/1506146.

5. Rupert Smith, *The Utility of Force: The Art of War in the Modern World* (London: Vintage, 2008).

6. Kurt M. Sutton, "A Look at the Marine Corps' Past," *Marines Magazine* (February 1997), http://www.geocities.com/heartland/6350/time.htm.

7. See, for example, Connable, "Culture Warriors"; Hoffman, " Origins and Application of the Expeditionary Ethos."

8. U.S. Marine Corps, *Small Wars Manual*, FMFRP 12–15 (Washington, D.C.: U.S. Government Printing Office, 1940).

9. Allison Abbe and Melissa Gouge, "Cultural Training for Military Personnel: Revisiting the Vietnam Era," *Military Review* (July–August 2012).

10. Michael E. Peterson, *The Combined Action Platoons: The US Marines' Other War in Vietnam* (New York: Praeger, 1989).

11. U.S. Department of Defense, "Defense Language Transformation Roadmap" (U.S. Department of Defense, May 2005).

12. U.S. Department of Defense, "DOD Regional and Cultural Capabilities: The Way Ahead" (Undersecretary of Defense, October 2007).

13. U.S. House of Representatives, "Building Language Skills and Cultural Competencies in the Military: DOD's Challenge in Today's Educational Environment" (Committee on Armed Services, November 2008).

14. U.S. House of Representatives, "Building Language Skills and Cultural Competencies," 9.

15. Allison Abbe, Lisa M. Gulick, and Jeffrey L. Herman, "Cross-Cultural Competence in Army Leaders: A Conceptual and Empirical Foundation" (U.S. Army Research Institute for the Behavioral and Social Sciences, October 2007); Daniel P. McDonald et al., "Developing and Managing Cross-Cultural Competence within the Department of Defense: Recommendations for Learning and Assessment" (Defense Language Office, October 2008).

16. U.S. Department of Defense, "Strategic Plan for Language Skills, Regional Expertise, and Cultural Capabilities, 2011–2016" (Department of Defense, 2011).

17. U.S. Department of Defense, "Strategic Plan for Language Skills."

18. U.S. Department of Defense, "Language, Regional Expertise and Cultural Capability (LREC) Identification, Planning and Sourcing, CJCSI 312.01A" (Chairman of the Joint Chiefs of Staff, January 31, 2013).

19. U.S. Government Accountability Office, "Military Training: Actions Needed to Improve Planning and Coordination of Army and Marine Corps Language and Culture Training" (U.S. Government Accountability Office, May 2011).

20. U.S. Government Accountability Office, "Military Training."

Chapter 6

1. Norman Cigar, *Al-Qaida, the Tribes and the Government: Lessons and Prospects for Iraq's Unstable Triangle*, Middle East Studies Occasional Papers No. 2 (Quantico, Va.: Marine Corps University Press, 2011); Timothy S. McWilliams and Kurtis P. Wheeler, *Al Anbar Awakening*, vol. 1, *American Perspectives U.S. Marines and Counterinsurgency in Iraq, 2004–2009* (Quantico, Va.: Marine Corps University Press, 2009); Neil Smith and Sean McFarland, "Anbar Awakens: The Tipping Point," *Military Review* (March–April 2008).

2. On a scale from 1 (not at all important) to 4 (very important), mean ranking of culture was 3.25 and of language was 3.09.

3. In order to better understand the culture and language perspectives of Marine Corps ground combat forces, a subset of 349 Marines with ground combat–related, two-digit Military Occupational Specialties (MOS 03, 08, 13, 18) was selected from the CAOCL survey. These ground combat respondents were compared to the remaining Marine Corps population in the survey (2,057 other respondents; total survey 2,406). Ninety-four percent of the ground combat arms (GCA) respondents had been deployed in the past four years, in contrast to 81 percent of the combat support (CS) respondents (p < .000).

4. T-test N = 1,997.

5. T-test GCA mean 2.65; versus CS respondents' mean 1.88 (1 = never; 4 = always) N = 2,455.

6. T-test GCA mean 3.11; versus CS respondents' mean 2.23 (1 = never; 4 = always) N = 2,455.

7. T-test GCA mean 3.31; versus CS respondents' mean 3.05 (1 = not important; 4 = very important) N = 2405.

8. T-test GCA mean 3.41; versus CS respondents' mean 3.22 (1 = not important; 4 = very important) N = 2,384.

9. On a scale from 1 (not at all important) to 4 (very important), mean ranking of culture was 3.25 and of language was 3.09. Paired-sample t-test N = 2384 (p < .000).

10. U.S. Marine Corps Recruit Depot (MCRD) Parris Island, "Command Brief" (2011).

11. At the time, he was the commanding officer of TECOM.

12. Twenty-four unarmed Iraqi civilians were killed by a group of Marines. All of the Marines were later acquitted.

13. An Iraqi girl was raped and murdered and her family killed by five U.S. soldiers—all of whom were found guilty.

14. Stanley Milgram, *Obedience to Authority: An Experimental View* (1974; repr., New York: Harper Perennial Modern Classics, 2009).

15. Philip Zimbardo, *The Lucifer Effect: Understanding How Good People Turn Evil* (New York: Random House, 2008).

Chapter 7

1. Elisabeth Bumiller, "We Have Met the Enemy and He Is PowerPoint," *New York Times*, April 26, 2010, http://www.nytimes.com/2010/04/27/world/27powerpoint.html?_r=1&ref=world.

2. Bumiller, "We Have Met the Enemy."

3. Dark Laughter, "Centcom Commander, 'Mad Dog' General James Mattis Set to Retire," *Duffel Blog* (April 27, 2012), http://www.duffelblog.com/2012/04/centcom-commander-general-mad-dog-mattis-set-to-retire/.

4. U.S. Marine Corps MAGTF Staff Training Program (MSTP), *Red Cell–Green Cell*, MSTP Pamphlet 2–0.1 (October 2011), Green-1.

5. In fact, the development of doctrine that accounts for the human aspects of the conflict in planning does represent a genuine adaptation in Marine Corps ways of conducting business. Since the Green Cell parallels familiar planning processes, this particular step can be seen as more like the "processing" of culture that I describe in the following chapter. However, since SMEs are usually hired to "translate" the cultural aspects of the problem to the planners, the Green Cell is discussed here, illustrating the overlapping nature of each of the four processes of adaptation for the Marine Corps described in Part II.

6. U.S. Army Headquarters and U.S. Marine Corps Combat Development Command, "Operational Terms and Graphics. FM 1–02 (FM 101–5–1) and MCRP 5–12a" (2010).

7. U.S. Human Terrain System, "Human Terrain Team Handbook" (Fort Leavenworth, Kans., September 2008), 2.

8. AAA Commission on the Engagement of Anthropology with the U.S. Security and Intelligence Communities (CEAUSSIC), "Final Report"; Patricia Cohen, "Panel Criticizes Military's Use of Embedded Anthropologists," *New York Times*, December 4, 2009; Carolyn Fluehr-Lobban, "Anthropology and Ethics in America's Declining Imperial Age," *Anthropology Today* 24, no. 4 (2008); Roberto J. Gonzalez, "Towards Mercenary Anthropology? The New US Army Counterinsurgency Manual FM 3–24 and the Military-Anthropology Complex," *Anthropology Today* 23, no. 3 (2007); Roberto J. Gonzalez, "Human Terrain," *Anthropology Today* 24, no. 1 (2008); Hugh Gusterson, "The U.S. Military's Quest to Weaponize Culture," *Bulletin of the Atomic Scientists* (June 20, 2008), http://www.thebulletin.org/print/web-edition/columnists/hugh-gusterson/the-us-militarys-quest-to-weaponize-culture; Vergano and Weise, "Should Anthropologists Work Alongside Soldiers?"

9. AAA Commission on the Engagement of Anthropology with the U.S. Security and Intelligence Communities (CEAUSSIC), "Final Report."

10. U.S. Human Terrain System, "Human Terrain Team Handbook," 11.

11. For example, in 2008 DoD announced its Minerva initiative, intended to attract social science research projects from civilian professors and universities. Topics in the following funding years have included anything from terrorism to religious extremism. See "2nd Annual Minerva Conference Minerva Research Summaries," September 15–16, 2011, http://minerva.dtic.mil/doc/MinervaResources2011.pdf. During the past six years, the Army Research Institute (ARI) expanded its existing staff of research social scientists and began conducting a number of projects focused on cultural and social issues. Similarly, the Office of Naval Research (ONR) broadened its traditional research funding (which had primarily emphasized technology and the "hard" sciences) to include a branch for social science research projects. (Since the Marine Corps comes under the umbrella of the Navy in terms of research funding, many of the studies related to the Marine Corps are funded by ONR.) While this list indicates some of the larger research funding ventures, numerous other social science

research projects were also contracted out independently by specific organizations within the military—far too many to list here.

Chapter 8

1. "Civilian" is a term used in the U.S. government and military to distinguish between nonmilitary personnel and employees who are military, must wear a uniform, and have signed a legal document permitting the government to send them wherever necessary. Government civilians are employed by the government or military but do not wear a uniform and are not required to move or deploy to other locations or combat zones (though a number do out of choice). Contractors, who are also civilians, work for private contracting firms but are sent to work for the military in the United States or overseas according to the contractual arrangements made between the government and these private firms.

2. See Barak Salmoni and Paula Holmes-Eber, *Operational Culture for the Warfighter: Principles and Applications* (Quantico, Va.: Marine Corps University, 2011).

3. "In-country" refers to events and activities held within a deployment area or country, in contrast to activities held in the United States (CONUS—Continental United States).

4. Group W, "Analytic Tools for the Application of Operational Culture: A Case Study in the Trans-Sahel" (Quantico. Va.: U.S. Marine Corps Combat Development Command, 2011).

5. U.S. Marine Corps, "Center for Advanced Operational Culture Learning Center of Excellence Charter (CAOCL COE)" (Quantico Va.: Marine Corps Combat Development Command, January 14, 2006).

6. Since I do not hold a clearance for ethical reasons, I never studied or followed the trajectory of MCIA, and so its role in this history is not included here.

7. "Sequestration" refers to the permanent reduction of government expenditures by a uniform percentage applied to all government services equally. In March 2013, due to the inability of Congress to ratify an alternative budget, sequestration went into effect, dramatically reducing the budgets of both civilian and military government programs.

8. ETP 150 literally means "Employ tactical phrases in 150 words (or less)," and focuses on providing a set of key memorized phrases that Marines can read or recite in the appropriate situations.

9. Budget cuts due to the U.S. government sequestration in 2013 led to a serious reduction and probable permanent cancellation of the program.

10. "Cordon and knock" literally refers to the process of cordoning (roping) off an area and then going from house to house, knocking on doors and then searching each one systematically.

11. ASCOPE stands for Area, Structures, Capabilities, Organizations, People and Events; PMESII stands for Political, Military, Economic, Social, Infrastructure and Information systems; and DIME represents Diplomatic, Informational, Military and/

or Economic aspects of the human environment. All of these acronyms represent popular Army templates for assessing the local human aspects of a conflict area.

12. Salmoni and Holmes-Eber, *Operational Culture for the Warfighter.*

13. It is important to note that MCIA developed a different approach to culture that focused much more deeply on analysis and critical thinking. Reflecting this focus, MCIA produced C-GIRH: Marine Corps Intelligence Activity, "C-GIRH Cultural Generic Information Requirements Handbook" (August 2008).

14. For all training requirements in the Marine Corps, there is an associated training and readiness manual, which describes the standards for what training objectives are to be met and how.

Chapter 9

1. Based on Pearson chi square test. N = 2,406.

2. U.S. Marine Corps, "Language, Regional and Culture Strategy 2011–2015" (2011).

3. The cost of a FAO is well over $200,000 per FAO when calculating his or her salary as well as education during the two to three years "away from" the Corps.

4. U.S. Department of the Navy, "Marine Corps Order 1250.11F" (Headquarters United States Marine Corps, March 27, 2013).

5. Marines, "U.S. Marine Corps Special Operations Command," http://www.marsoc.marines.mil/About.aspx.

6. U.S. Marine Corps Force Structure Review Group, "Reshaping America's Expeditionary Force"; U.S. Marine Corps, "Vision and Strategy 2025"; U.S. Marine Corps, "Long War."

7. U.S. Marine Corps Force Structure Review Group, "Reshaping America's Expeditionary Force."

8. U.S. Marine Corps Force Structure Review Group, "Reshaping America's Expeditionary Force," 4.

9. U.S. Marine Corps, "The Marine Corps Regional, Culture, and Language Familiarization Program" (August 2011).

10. U.S. Marine Corps, "Marine Corps Regional, Culture, and Language Familiarization Program."

11. The U.S. military splits the world up into six commands: NORTHCOM (North America), SOUTHCOM (South America), AFRICOM (Africa), PACOM (Pacific), CENTCOM (Middle East), and EUCOM (Europe).

12. Anthony Terlizzi, "Cross-Cultural Competence and Civil-Military Operations (CMO)," in *Cross-Cultural Competence for a Twenty-First Century Military: Culture the Flipside of the COIN,* ed Robert Greene Sands and Allison Greene-Sands (Plymouth, U.K. Lexington Books, 2014).

13. U.S. Marine Corps, "Marine Corps Regional, Culture, and Language Familiarization Program."

Conclusion

1. Simons, *Company They Keep.*
2. Fujimura, "Culture in/Culture of the United States Naval Academy"; Fujimura, "Living the Dream."
3. Tortorello, "Ethnography of 'Courage' among U.S. Marines."
4. Ben-Ari, *Mastering Soldiers.*
5. Russell, *Innovation, Transformation, and War.*
6. Yanow, *How Does a Policy Mean?*

Bibliography

"A Basic Guide to Working with the United States Marine Corps." Unpublished manuscript.

"Marines Deliver Lifesaving Supplies to Flood Stricken Pakistanis." *Marine Corps Gazette* (January 9, 2013). http://www.mca-marines.org/gazette/marines-deliver-lifesaving-supplies-flood-stricken-pakistanis.

"Red Cross Pulls Out of Baghdad, Basra." *ABC News* (November 8, 2003). http://www.abc.net.au/news/2003–11–08/red-cross-pulls-out-of-baghdad-basra/1506146.

"Truck Bomb Kills Chief U.N. Envoy to Iraq: 17 Dead, 100 Injured by Explosion at U.N. Headquarters." *CNN* (August 20, 2003). http://www.cnn.com/2003/WORLD/meast/08/19/sprj.irq.main/.

"26th MEU Slated to Deploy to Afghanistan." *Marine Corps Gazette* (2011). http://www.mca-marines.org/gazette/26th-meu-slated-deploy-afghanistan.

"US Soldiers Killed in Baghdad." *BBC News* (July 7, 2003). http://news.bbc.co.uk/2/hi/middle_east/3050358.stm.

AAA Commission on the Engagement of Anthropology with the U.S. Security and Intelligence Communities (CEAUSSIC). "Final Report on the Army's Human Terrain System Proof of Concept Program." American Anthropological Association, October 14, 2009.

Abbe, Allison, and Melissa Gouge. "Cultural Training for Military Personnel: Revisiting the Vietnam Era." *Military Review* (July–August 2012): 9–17.

Abbe, Allison, Lisa M. Gulick, and Jeffrey L. Herman. "Cross-Cultural Competence in Army Leaders: A Conceptual and Empirical Foundation." U.S. Army Research Institute for the Behavioral and Social Sciences, October 2007.

Abu Lughod, Lila. *Veiled Sentiments: Honor and Poetry in a Bedouin Society.* Berkeley: University of California Press, 1986.

Adamsky, Dima. *The Culture of Military Innovation: The Impact of Cultural Factors on the Revolution in Military Affairs in Russia, the US, and Israel.* Stanford, Calif.: Stanford Security Studies, 2010.

Albro, Robert. "Public Anthropology and Multitrack Dialoguing in the Securityscape." In *Anthropologists in the Securityscape: Ethics, Practice and Professional Identity,* ed. Robert Albro, George E. Marcus, Laura A. McNamara, and Monica Schoch-Spana, 39–56. Walnut Creek, Calif.: Left Coast Press, 2012.

Albro, Robert, George E. Marcus, Laura A. McNamara, and Monica Schoch-Spana, eds. *Anthropologists in the Securityscape: Ethics, Practice and Professional Identity.* Walnut Creek, Calif.: Left Coast Press, 2012.

Amos, James F. "Role of the United States Marine Corps." Memorandum for the Secretary of Defense from the Commandant of the Marine Corps, September 12, 2011.

———. "Who We Are." *Proceedings* 138, no. 11 (2012): 17–21.

Bar, Neta, and Eyal Ben-Ari. "Israeli Snipers in the Al-Aqsa Intifada: Killing, Humanity and Lived Experience." *Third World Quarterly* 26, no. 1 (2005): 137–56.

Belkin, Aaron. *Bring Me Men: Military Masculinity and the Benign Facade of American Empire, 1898–2001.* New York: Columbia University Press, 2012.

Ben-Ari, Eyal. "Masks and Soldiering: The Israeli Army and the Palestinian Uprising." *Cultural Anthropology* 4, no. 4 (1989): 372–89.

———. *Mastering Soldiers: Conflict, Emotions and the Enemy in an Israeli Military Unit.* Oxford: Bergham Books, 1998.

Ben-Ari, Eyal, and Sabine Fruhstuck. "The Celebration of Violence: A Live-Fire Demonstration Carried Out by Japan's Contemporary Military." *American Ethnologist* 30, no. 4 (2003): 540–55.

Birkland, Thomas A. *An Introduction to the Policy Process: Theories, Concepts and Models of Public Policy Making.* 3rd ed. Armonk, N.Y.: M. E. Sharpe, 2011.

Black, Jeremy. *War and the Cultural Turn.* Cambridge: Polity, 2010.

Bouhan, Neil, and Paul Swartz. "Trends in U.S. Military Spending." Maurice Greenberg Center for Geoeconomic Studies, Council on Foreign Relations, June 28, 2011.

Bourdieu, Pierre. *The Logic of Practice.* Trans. Richard Nice. Stanford, Calif.: Stanford University Press, 1992.

———. *Outline of a Theory of Practice.* Trans. Richard Nice. Cambridge: Cambridge University Press, 1977.

Brenneis, Don. "A Partial View of Contemporary Anthropology." *American Anthropologist* 106, no. 3 (2004): 580–88.

Britt, Thomas W., Amy B. Adler, and Carl A. Castro, eds. *Military Life: The Psychology of Serving in Peace and Combat.* Vol. 4. Westport, Conn.: Praeger Security International, 2006.

Brown, Keith. "'All They Understand Is Force': Debating Culture in Operation Iraqi Freedom." *American Anthropologist* 110, no. 4 (2008): 443–53.

Brown, Melissa T. *Enlisting Masculinity: The Construction of Gender in US Military Recruiting Advertising during the All-Volunteer Force.* Oxford Studies in Gender and International Relations. Oxford: Oxford University Press, 2012.

Buley, Benjamin. *The New American Way of War: Military Culture and the Political Utility of Force*. London: Routledge, 2008.

Bumiller, Elisabeth. "We Have Met the Enemy and He Is PowerPoint." *New York Times*, April 26, 2010. http://www.nytimes.com/2010/04/27/world/27powerpoint.html?_r=1&ref=world.

Burk, James, ed. *The Adaptive Military: Armed Forces in a Turbulent World*. Boulder, Colo.: Westview Press, 1994.

———, ed. *How 9/11 Changed our Ways of War*. Stanford, CA: Stanford University Press, 2013.

Burke, Carol. *Camp All American, Hanoi Jane and the High-and-Tight: Gender, Folklore and Changing Military Culture*. Boston: Beacon Press, 2005.

Caforio, Guiseppe, ed. *Handbook of the Sociology of the Military*. New York: Springer, 2006.

Cassidy, Robert M. *Counterinsurgency and the Global War on Terror: Military Culture and Irregular War*. Stanford, Calif.: Stanford Security Studies, 2008.

Castro, Celso. "Anthropological Methods and the Study of the Military: The Brazilian Experience." In *Qualitative Methods in Military Studies*, ed. Helena Carreiras and Celso Castro, 8–17. London: Routledge, 2012.

Chivers, C. J. "A Grueling Course for Training Marine Officers Will Open Its Doors to Women." *New York Times*, July 8, 2012.

Chomsky, Noam, and Misou Rinat. *On Language: Chomsky's Classic Works Language and Responsibility and Reflections on Language in One Volume*. New York: New Press, 1998.

Cigar, Norman. *Al-Qaida, the Tribes and the Government: Lessons and Prospects for Iraq's Unstable Triangle*. Middle East Studies Occasional Papers No. 2. Quantico, Va.: Marine Corps University Press, 2011.

Clark, Gracia. "Working the Field: Kumasi Central Market as Community, Employer and Home." *Anthropology of Work Review* 29, no. 3 (2008): 69–75.

Cohen, Eliot A. "Change and Transformation in Military Affairs." *Journal of Strategic Studies* 27, no. 3 (2004): 395–407.

Cohen, Patricia. "Panel Criticizes Military's Use of Embedded Anthropologists." *New York Times*, December 4, 2009.

Coker, Christopher. *The Warrior Ethos: Military Culture and the War on Terror*. New York: Routledge, 2007.

Colon, Santiago G. "Living on Ship: 26th MEU Marines Balance Duty and Downtime." www.26thmeu.marines.mil/News/NewsArticleDisplay/tabid/2723/Article/65684/living-on-ship-26th-meu-marines-balance-duty-and-downtime.aspx.

Connable, Ben. "Culture Warriors: Marine Corps Organizational Culture and Adaptation to Cultural Terrain." *Small Wars Journal* (2008). http://smallwarsjournal.com/jrnl/art/culture-warriors.

D'Andrade, Roy. "Some Methods for Studying Cultural Cognitive Structures." In *Finding Culture in Talk: A Collection of Methods*, ed. Naomi Quinn, 83–104. New York: Palgrave-Macmillan, 2005.

Davis, Rochelle. "Culture as a Weapon System." *Middle East Report* 255 (Summer 2010): 8–13.

De Angelis, Karin, and David R. Segal. "Minorities in the Military." In *The Oxford Handbook of Military Psychology*, ed. Janice H. Laurence and Michael D. Matthews, 325–43. Oxford: Oxford University Press, 2011.

Desai, Gaurav Gajanan. *Postcolonialisms: An Anthology of Cultural Theory and Criticism*. New Brunswick, N.J.: Rutgers University Press, 2005.

Dorn, Edwin, Howard D. Graves, and Walter F. Ulmer. *American Military Culture in the Twenty-First Century: A Report of the CSIS International Security Program*. Washington, D.C.: Center for Strategic and International Studies, 2000.

Dunn, William N. *Public Policy Analysis*. 5th ed. Boston: Pearson, 2012.

Edwards, David B. "Counterinsurgency as a Cultural System." *Small Wars Journal* (2010). http://smallwarsjournal.com/jrnl/art/counterinsurgency-as-a-cultural-system.

English, Allan D. *Understanding Military Culture: A Canadian Perspective*. Montreal: McGill-Queen's University Press, 2004.

Estrada, Armando X. "Gay Service Personnel in the U.S. Military: History, Progress and a Way Forward." In *The Oxford Handbook of Military Psychology*, ed. Janice H. Laurence and Michael D. Matthews, 344–64. Oxford: Oxford University Press, 2011.

FallenUSSoldiers. "U.S. Marine Corps—Making a Marine." http://www.youtube.com/watch?v=MYRccSZgXV4.

———. "U.S. Marine Corps—Making a Marine Part 2." http://www.youtube.com/watch?v=aPgkACv7grk&feature=mfu_in_order&list=UL.

———. "U.S. Marine Corps—Making a Marine Part 3." http://www.youtube.com/watch?v=fPRzONdtgNo.

Farrell, Theo. "Figuring Out Fighting Organisations: The New Organisational Analysis in Strategic Studies." *Journal of Strategic Studies* 19, no. 1 (1996): 122–35.

———. *The Norms of War: Cultural Beliefs and Modern Conflict*. Boulder, Colo.: Lynne Rienner, 2005.

Farrell, Theo, and Terry Terriff, eds. *The Sources of Military Change: Culture, Politics, Technology*. Boulder, Colo.: Lynne Rienner, 2001.

Farrell, Theo, Terry Terriff, and Osinga Frans, eds. *A Transformation Gap: American Innovations and European Military Change*. Stanford, Calif.: Stanford Security Studies, 2010.

Fick, Nathaniel C. *One Bullet Away: The Making of a Marine Officer*. New York: Mariner Books, 2006.

Fluehr-Lobban, Carolyn. "Anthropology and Ethics in America's Declining Imperial Age." *Anthropology Today* 24, no. 4 (2008): 18–22.

Fosher, Kerry. *Making Homeland Security at the Local Level*. Chicago: University of Chicago Press, 2009.

Frese, Pamela R. "Guardians of the Golden Age: Custodians of U.S. Military Culture." In *Anthropology and the United States Military: Coming of Age in the Twenty-First*

Century, ed. Pamela R. Frese and Margaret C. Harrell, 45–68. New York: Palgrave-Macmillan, 2003.

Frese, Pamela R., and Margaret C. Harrell, eds. *Anthropology and the United States Military: Coming of Age in the Twenty-First Century*. New York: Palgrave-Macmillan, 2003.

Frost, Peter J., Larry F. Moore, Meryl R. Louis, Craig C. Lundberg, and Joanne Martin, eds. *Reframing Organizational Culture*. Newbury Park, Calif.: Sage, 1991.

Fujimura, Clementine. "Culture in/Culture of the United States Naval Academy." In *Anthropologists in the Securityscape: Ethics, Practice and Professional Identity*, ed. Robert Albro, George E. Marcus, Laura A. McNamara, and Monica Schoch-Spana, 115–28. Walnut Creek, Calif.: Left Coast Press, 2012.

———. "'Living the Dream': One Military Anthropologist's Initiation " In *Practicing Military Anthropology: Beyond Expectations and Traditional Boundaries*, ed. Robert Rubinstein, Kerry Fosher, and Clementine Fujimura, 29–44. Sterling, Va.: Kumarian Press, 2013.

Geertz, Clifford. *The Interpretation of Cultures*. New York: Basic Books, 1977.

Gonzalez, Roberto J. *American Counterinsurgency: Human Science and the Human Terrain*. Chicago: Prickly Paradigm Press, 2009.

———. "Human Terrain." *Anthropology Today* 24, no. 1 (2008): 21–26.

———. "Towards Mercenary Anthropology? The New US Army Counterinsurgency Manual FM 3–24 and the Military-Anthropology Complex." *Anthropology Today* 23, no. 3 (2007): 14–19.

Group W. "Analytic Tools for the Application of Operational Culture: A Case Study in the Trans-Sahel." Quantico, Va.: U.S. Marine Corps Combat Development Command, 2011.

Gusterson, Hugh. "Anthropology and Militarism." *Annual Review of Anthropology* 36 (2007): 155–75.

———. "Anthropology and the Military: 1968, 2003 and Beyond?" *Anthropology Today* 19, no. 3 (2003): 25–26.

———. *Nuclear Rites: A Weapons Laboratory at the End of the Cold War*. Berkeley: University of California Press, 1998.

———. "Studying up Revisited." *PoLAR: Political and Legal Anthropology Review* 20, no. 1 (1997): 114–19.

———. "The U.S. Military's Quest to Weaponize Culture." *Bulletin of the Atomic Scientists* (June 20, 2008). http://www.thebulletin.org/print/web-edition/columnists/hugh-gusterson/the-us-militarys-quest-to-weaponize-culture.

Hamada, Tomoko, and Willis E. Sibley, eds. *Anthropological Perspectives on Organizational Culture*. Lanham, Md.: University Press of America, 1994.

Hanks, William F. "Pierre Bourdieu and the Practices of Language." *Annual Review of Anthropology* 34 (2005): 67–83.

Harrell, Margaret C. "Gender and Class-Based Role Expectations for Army Spouses." In *Anthropology and the United States Military: Coming of Age in the Twenty-First*

Century, ed. Pamela R. Frese and Margaret C. Harrell, 69–94. New York: Palgrave-Macmillan, 2003.

Hawkins, John P. *Army of Hope, Army of Alienation: Culture and Contradiction in the American Army Communities of Cold War Germany.* Westport, Conn.: Praeger, 2001.

Herek, Gregory M., and Aaron Belkin. "Sexual Orientation and Military Service: Prospects for Organizational and Institutional Change in the United States." In *Military Life: The Psychology of Serving in Peace and Combat*, ed. Thomas W. Britt, Amy B. Adler, and Carl Andrew Castro, 119–42. Westport, Conn.: Praeger Security International, 2006.

Hill, Jane H. "Finding Culture in Narrative." In *Finding Culture in Talk: A Collection of Methods*, ed. Naomi Quinn, 157–202. New York: Palgrave-Macmillan, 2005.

Hill, Michael, and Peter Hupe. *Implementing Public Policy: An Introduction to the Study of Operational Governance.* 2nd ed. Thousand Oaks, Calif.: Sage, 2008.

Hoffman, Frank. "The Origins and Application of the Expeditionary Ethos." Washington, D.C.: Scitor, 2007.

Hofstede, Geert. *Cultures and Organizations: Software of the Mind.* 3rd ed. New York: McGraw Hill, 2010.

Holmes-Eber, Paula. *Daughters of Tunis: Women, Family and Networks in a Muslim City.* Boulder, Colo.: Westview Press, 2003.

———. "A Day in the Life of the Marine Corps Professor of Operational Culture." In *Practicing Military Anthropology: Beyond Expectations and Traditional Boundaries*, ed. Robert Rubinstein, Kerry Fosher, and Clementine Fujimura, 45–64. Sterling, Va.: Kumarian Press, 2013.

———. "Teaching Culture at Marine Corps University." In *Anthropologists in the Securityscape: Ethics, Practice and Professional Identity*, ed. Robert Albro, George E. Marcus, Laura A. McNamara, and Monica Schoch-Spana, 129–42. Walnut Creek, Calif.: Left Coast Press, 2012.

Hunt, James G., and Robert L. Phillips. "Leadership in Battle and Garrison: A Framework for Understanding the Differences and Preparing for Both." In *Handbook of Military Psychology*, ed. Reuven Gal and A. David Mangelsdorff, 393–410. Chichester: John Wiley and Sons, 1991.

Ingram, Kate. "Equal Opportunities, the Clothing Industry and the Law." In *Culture and Policy in Northern Ireland: Anthropology in the Public Arena*, ed. Hastings Donnan and Graham Farlane, 153–66. Chester Springs, Pa.: Dufour Editions, 1997.

Jones, Michael Owen. *Studying Organizational Symbolism: What, How, Why?* Thousand Oaks, Calif.: Sage, 1996.

Karatzogianni, Athina, ed. *Violence and War in Culture and the Media: Five Disciplinary Lenses.* London: Routledge, 2012.

Kelly, John D., Beatrice Jauregui, Sean T. Mitchell, and Jeremy Walton, eds. *Anthropology and Global Counterinsurgency.* Chicago: University of Chicago Press, 2010.

Keyton, JoAnne. *Communication and Organizational Culture: A Key to Understanding Work Experiences.* Thousand Oaks, Calif.: Sage, 2010.

Killcullen, David. *The Accidental Guerrilla*. Oxford: Oxford University Press, 2009.

Kingdon, John. *Agendas, Alternatives and Public Policies*. 2nd ed. New York: Longman, 2003.

Kirke, Charles. "Insider Anthropology: Theoretical and Empirical Issues for the Researcher." In *Qualitative Methods in Military Studies*, ed. Helena Carreiras and Celso Castro, 17–30. London: Routledge, 2012.

Kraft, Michael E. *Public Policy: Politics, Analysis and Alternatives*. 3rd ed. Washington, D.C.: CQ Press, 2009.

Krulak, Victor H. *First to Fight: An Inside View of the Marine Corps*. Annapolis, Md.: U.S. Naval Institute Press, 1989.

Kryou, Christos, and Robert Rubinstein. "Cultural Anthropology Studies of Conflict." In *Encyclopedia of Violence, Peace and Conflict*, ed. Lester Kurtz, 515–21. Amsterdam: Elsevier, 2008.

Labuc, Stasiu. "Cultural and Societal Factors in Military Organizations." In *Handbook of Military Psychology*, ed. Reuven Gal and A. David Mangelsdorff, 471–90. Chichester: John Wiley and Sons, 1991.

Laurence, Janice H., and Michael D. Matthews, eds. *The Oxford Handbook of Military Psychology*. Oxford: Oxford University Press, 2011.

Levi-Strauss, Claude. *The Origin of Table Manners*. New York: Harper and Row, 1968.

———. *The Savage Mind*. Trans. George Weidenfeld. Chicago: University of Chicago Press, 1962.

Lierner, Piero C. "Side Effects of the Chain of Command on Anthropological Research: The Brazilian Army." In *Qualitative Methods in Military Studies*, ed. Helena Carreiras and Celso Castro, 68–84. London: Routledge, 2012.

Lipsky, Michael. *Street Level Bureaucracy: Dilemmas of the Individual in Public Service*. 30th anniversary edition. 1980. Reprint, New York: Russell Sage Foundation, 2010.

Lucas, George R., ed. *Anthropologists in Arms: The Ethics of Military Anthropology*. Walnut Creek, Calif.: AltaMira Press, 2009.

Luft, Gal. *Beer, Bacon and Bullets: Culture in Coalition Warfare from Gallipoli to Iraq*. Charleston, S.C.: BookSurge, 2010.

Lutz, Catherine, ed. *The Bases of Empire: The Global Struggle against Military Posts*. New York: NYU Press, 2009.

———. *Homefront: A Military City and the American Twentieth Century*. Boston: Beacon Press, 2002.

Mahnken, Thomas. *Technology and the American Way of War since 1945*. New York: Columbia University Press, 2010.

Marine OCS. "Marine Corps Commercial: Towards the Sounds of Chaos." http://www.youtube.com/watch?v=tYrBSTBHCS4.

Marines. "Marine Corps Commercial: A Path for Warriors." http://www.youtube.com/watch?v=S_nApy9VVlk.

———. "U.S. Marine Corps Special Operations Command." http://www.marsoc.marines.mil/About.aspx.

Martin, Joanne. *Cultures in Organizations: Three Perspectives*. Oxford: Oxford University Press, 1992.

———. *Organizational Culture: Mapping the Terrain*. Thousand Oaks, Calif.: Sage, 2001.

McDonald, Daniel P., Gary McGuire, Joan Johnston, Brian Selmeski, and Allison Abbe. "Developing and Managing Cross-Cultural Competence within the Department of Defense: Recommendations for Learning and Assessment." Defense Language Office, October 2008.

McFate, Montgomery. "Anthropology and Counterinsurgency: The Strange Story of Their Curious Relationship." *Military Review* (March–April 2005): 24–38.

McFate, Montgomery, Britt Damon, and Robert Holiday. "What Do Commanders Really Want to Know? US Army Human Terrain System Lessons Learned from Iraq and Afghanistan." In *The Oxford Handbook of Military Psychology*, ed. Janice H. Laurence and Michael D. Matthews, 92–113. Oxford: Oxford University Press, 2011.

McWilliams, Timothy S., and Kurtis P. Wheeler. *Al Anbar Awakening*. Vol. 1, *American Perspectives U.S. Marines and Counterinsurgency in Iraq, 2004–2009*. Quantico, Va.: Marine Corps University Press, 2009.

Milgram, Stanley. *Obedience to Authority: An Experimental View*. 1974. Reprint, New York: Harper Perennial Modern Classics, 2009.

Moskos, Charlie. "From Institution to Occupation: Trends in Military Organizations." *Armed Forces and Society* 4, no. 1 (1977): 41–50.

———. "Institutional and Occupational Trends in Armed Forces: An Update." *Armed Forces and Society* 12, no. 3 (1986): 377–82.

Moskos, Charlie, and James Burk. "The Postmodern Military." In *The Adaptive Military: Armed Forces in a Turbulent World*, ed. James Burk, 163–82. Boulder, Colo.: Westview Press, 1994.

Moskos, Charlie, John A. Williams, and David R. Segal, eds. *The Postmodern Military*. Oxford: Oxford University Press, 2000.

Nader, Laura. "Up the Anthropologist—Perspectives Gained from Studying Up." In *Reinventing Anthropology*, ed. Dell H. Hymes, 284–311. New York: Pantheon Books, 1972.

Nagl, John A. *Learning to Eat Soup with a Knife: Counterinsurgency Lessons from Malaya and Vietnam*. Chicago: University of Chicago Press, 2005.

Navarro, Alejandra. "Negotiating Access to an Argentinean Military Institution in Democratic Times: Difficulties and Challenges." In *Qualitative Methods in Military Studies*, ed. Helena Carreiras and Celso Castro, 85–96. London: Routledge, 2012.

Network of Concerned Anthropologists. *The Counter-Counterinsurgency Manual*. Chicago: Prickly Paradigm Press, 2009.

Nuciari, Marina. "Models and Explanations for Military Organization: An Updated Reconsideration." In *Handbook of the Sociology of the Military*, ed. Guiseppe Caforio, 61–85. New York: Springer, 2006.

Office of the Undersecretary of Defense. "National Defense Budget Estimates for FY 2014." Comptroller, May 2013. http://comptroller.defense.gov/defbudget/fy2014/FY14_Green_Book.pdf.

Okongwu, Anne Francis, and Joan P. Mencher. "The Anthropology of Public Policy: Shifting Terrains." *Annual Review of Anthropology* 29 (2000): 107–24.

Otterbein, Keith F. *The Anthropology of War.* Long Grove, Ill.: Waveland Press, 2009.

OurMarines. "'No Compromises': Marine Commercial." http://www.youtube.com/watch?v=opAJtwyujvo.

Peterson, Michael E. *The Combined Action Platoons: The US Marines' Other War in Vietnam.* New York: Praeger, 1989.

Petronio, Katie. "Get Over It! We Are Not Created Equal." *Marine Corps Gazette* (2012). http://www.mca-marines.org/gazette/article/get-over-it-we-are-not-all-created-equal.

Pierce, James G. *Is the Organizational Culture of the U.S. Army Congruent with the Professional Development of Its Senior Level Officer Corps?* Carlisle, Pa.: Strategic Studies Institute, U.S. Army War College, 2010.

Pierce, Penny F. "The Role of Women in the Military." In *Military Life: The Psychology of Serving in Peace and Combat,* ed. Thomas W. Britt, Amy B. Adler, and Carl Andrew Castro, 97–118. Westport, Conn.: Praeger Security International, 2006.

Porter, Patrick. *Military Orientalism: Eastern War through Western Eyes.* New York: Columbia University Press, 2009.

Prentice, Rebecca. "Knowledge, Skill and the Inculcation of the Anthropologist: Reflections on Learning to Sew in the Field." *Anthropology of Work Review* 29, no. 3 (2008): 54–61.

Pressman, Jeffrey L., and Aaron Wildavsky. *Implementation: How Great Expectations in Washington Are Dashed in Oakland; or Why It's Amazing That Federal Programs Work at All.* Berkeley: University of California Press, 1984.

Price, David H. *Anthropological Intelligence: The Deployment and Neglect of American Anthropology in the Second World War.* Durham, N.C.: Duke University Press, 2008.

Quinn, Naomi. "How to Reconstruct Schemas People Share from What They Say." In *Finding Culture in Talk: A Collection of Methods,* ed. Naomi Quinn, 35–82. New York: Palgrave-Macmillan, 2005.

Richards, Paul, ed. *No Peace No War: An Anthology of Contemporary Armed Conflicts.* Oxford: James Currey, 2005.

Ricks, Thomas E. *Making the Corps.* New York: Scribner, 2007.

Robben, Antonius C. G. M., ed. *Iraq at a Distance: What Anthropologists Can Teach Us About the War.* Philadelphia: University of Pennsylvania Press, 2010.

Robertson, Jennifer. "Reflexivity Redux: A Pithy Polemic on 'Positionality.'" *Anthropological Quarterly* 75, no. 4 (2002): 785–92.

Rosen, Stephen P. "Military Effectiveness: Why Society Matters." *International Security* 19, no. 4 (1995): 5–31.

————, ed. *Winning the Next War: Innovation and the Modern Military*. Cornell Studies in Security Affairs. Ithaca, N.Y.: Cornell University Press, 1994.

Rosenberg, Howard L. "Real Story of Jessica Lynch's Convoy." *ABC News Nightline* (June 17, 2003). http://abcnews.go.com/Nightline/story?id=128387&page=1.

Rubinstein, Robert. "Master Narratives, Retrospective Attribution and Ritual Pollution in Anthropology's Military Engagements with the Military." In *Practicing Military Anthropology: Beyond Expectations and Traditional Boundaries*, ed. Robert Rubinstein, Kerry Fosher, and Clementine Fujimura, 119–35. Sterling, Va.: Kumarian Press, 2013.

————. *Peacekeeping under Fire: Culture and Intervention*. Boulder, Colo.: Paradigm, 2008.

Rubinstein, Robert, Kerry Fosher, and Clementine Fujimura, eds. *Practicing Military Anthropology: Beyond Expectations and Traditional Boundaries*. Sterling, Va.: Kumarian Press, 2013.

Russell, James A. *Innovation, Transformation, and War: Counterinsurgency Operations in Anbar and Ninewa Provinces, Iraq, 2005–2007*. Stanford, Calif.: Stanford Security Studies, 2010.

Sabatier, Paul, ed. *Theories of the Policy Process*. Boulder, Colo.: Westview Press, 2007.

Sahlins, Marshall. *Culture in Practice: Selected Essays*. New York: Zone Books, 2000.

Salmoni, Barak, and Paula Holmes-Eber. *Operational Culture for the Warfighter: Principles and Applications*. Quantico, Va.: Marine Corps University, 2011.

Sarkesian, Sam C., John Allen Williams, and Stephen J. Cimbala. *U.S. National Security: Policymakers, Processes and Politics*. 4th ed. Boulder, Colo.: Lynne Rienner, 2008.

Schein, Edgar H. *Organizational Culture and Leadership*. 3rd ed. San Francisco: Jossey-Bass, 2004.

Schultz, Majken. *On Studying Organizational Cultures: Diagnosis and Understanding*. Berlin: Walter de Gruyter, 1995.

Schwartzman, Helen. *Ethnography in Organizations*. Newbury Park, Calif.: Sage, 1992.

————. *The Meeting*. New York: Plenum Press, 1989.

Shimko, Keith. *The Iraq Wars and America's Military Revolution*. Cambridge: Cambridge University Press, 2010.

Shore, Chris, Susan Wright, and David Pero, eds. *Policy Worlds: Anthropology and the Analysis of Contemporary Power*. Oxford: Berghahn Books, 2011.

Siegel, Adam. "A Chronology of U.S. Marine Corps Humanitarian Assistance and Peace Operations." Alexandria, Va.: Center for Naval Analyses, 1994.

Simons, Anna. *The Company They Keep: Life Inside the U.S. Army Special Forces*. New York: Free Press, 1997.

————. "War: Back to the Future." *Annual Review of Anthropology* 28 (1999): 73–108.

Singal, Jesse, Christine Lim, and M. J. Stephey. "May 2003 Victory Lap." *Time*, March 19, 2010. http://www.time.com/time/specials/packages/article/0,28804,1967340_1967342_1967406,00.html.

Sluka, Jeffrey A. "Curiouser and Curiouser: Montgomery Mcfate's Strange

Interpretation of the Relationship between Anthropology and Counterinsurgency." *PoLAR: Political and Legal Anthropology Review* 33, no. s1 (2010): 99–115.

Smith, James M. "Service Cultures, Joint Cultures and the US Military." *Airman-Scholar* (Winter 1998): 120–35.

Smith, Kevin B., and Christopher W. Larimer. *The Public Policy Theory Primer.* Boulder, Colo.: Westview Press, 2009.

Smith, Neil, and Sean McFarland. "Anbar Awakens: The Tipping Point," *Military Review* (March–April 2008): 41-52.

Smith, Rupert. *The Utility of Force: The Art of War in the Modern World.* London: Vintage, 2008.

Soeters, Joseph L. "Culture in Uniformed Organizations." In *Handbook of Organizational Culture and Climate*, ed. Neal M. Ashkanasy, Celeste Wilderom, and Mark F. Peterson, 465–83. Thousand Oaks, Calif.: Sage, 2000.

Soeters, Joseph L., and R. Recht. "Culture and Discipline in Military Academies: An International Comparison." *Journal of Political and Military Sociology* 26, no. 2 (1998): 169–89.

Soeters, Joseph L., Cristina-Rodica Poponete, and Joseph T. Page Jr. "Culture's Consequences in the Military." In *Military Life: The Psychology of Serving in Peace and Combat*, ed. Thomas W. Britt, Amy B. Adler, and Carl Andrew Castro, 13–34. Westport, Conn.: Praeger Security International, 2006.

Soeters, Joseph L., Donna J. Winslow, and Alise Weibull. "Military Culture." In *Handbook of the Sociology of the Military*, ed. Guiseppe Caforio, 237–54. New York: Springer, 2006.

Sondhaus, Lawrence. *Strategic Studies and Ways of War.* London: Routledge, 2006.

Spradley, James P. *Participant Observation.* Belmont, Calif.: Wadsworth, 1980.

Stroeken, Koen, ed. *War, Technology, Anthropology.* New York: Berghahn Books, 2012.

Sturkey, Marion F. *Warrior Culture of the U.S. Marines: Axioms for Warriors, Marine Quotations, Battle History, Reflections on Combat, Corps Legacy, Humor—and Much More—for the World's Warrior Elite.* Plum Branch, S.C.: Heritage Press International, 2003.

Sutton, Kurt M. "A Look at the Marine Corps' Past." *Marines Magazine* (February 1997). http://www.geocities.com/heartland/6350/time.htm.

Terlizzi, Anthony. "Cross-Cultural Competence and Civil-Military Operations (CMO)." In *Cross-Cultural Competence for a Twenty-First Century Military: Culture the Flipside of the COIN*, ed Robert Greene Sands and Allison Greene-Sands, 131–140. Plymouth, U.K.: Lexington Books, 2014.

Terriff, Terry. "'Innovate or Die': Organizational Culture and the Origins of Maneuver Warfare in the United States Marine Corps." *Journal of Strategic Studies* 29, no. 3 (2006): 475–503.

———. "Warriors and Innovators: Military Change and Organizational Culture in the US Marine Corps." *Defence Studies* 6, no. 2 (2006): 215–47.

Tortorello, Frank J. "An Ethnography of 'Courage' among U.S. Marines." Ph.D. diss., University of Illinois at Urbana-Champaign, 2010.

U.S. Army and U.S. Marine Corps. *Counterinsurgency Field Manual. U.S. Army Field Manual No. 3–24. Marine Corps Warfighting Publication No. 3–33.5.* Chicago: University of Chicago Press, 2007.

U.S. Army Headquarters and U.S. Marine Corps Combat Development Command. "Operational Terms and Graphics. FM 1–02 (FM 101–5–1) and MCRP 5–12a." 2010.

U.S. Congress. "National Security Act (Unification Bill)." 1947 (amended 1952).

U.S. Department of Defense. "Defense Language Transformation Roadmap." U.S. Department of Defense, May 2005.

———. "DOD Regional and Cultural Capabilities: The Way Ahead." Undersecretary of Defense, October 2007.

———. "Language, Regional Expertise and Cultural Capability (LREC) Identification, Planning and Sourcing, CJCSI 312.01A." Chairman of the Joint Chiefs of Staff, January 31, 2013.

———. "Strategic Plan for Language Skills, Regional Expertise, and Cultural Capabilities, 2011–2016." Department of Defense, 2011.

U.S. Department of the Navy. "Marine Corps Order 1250.11F." Headquarters United States Marine Corps, March 27, 2013.

U.S. Government Accountability Office (GAO). "Military Training: Actions Needed to Improve Planning and Coordination of Army and Marine Corps Language and Culture Training." U.S. Government Accountability Office, May 2011.

U.S. House of Representatives. "Building Language Skills and Cultural Competencies in the Military: DOD's Challenge in Today's Educational Environment." Committee on Armed Services, November 2008.

U.S. Human Terrain System. "Human Terrain Team Handbook." Fort Leavenworth, Kans., September 2008.

U.S. Marine Corps. "Center for Advanced Operational Culture Learning Center of Excellence Charter (CAOCL COE)." Quantico, Va.: Marine Corps Combat Development Command, January 14, 2006.

———. "Commandant's Planning Guidance." General Amos, 35th Commandant of the Marine Corps.

———. "Concepts and Programs." Headquarters Marine Corps, 2010.

———. "Concepts and Programs." Headquarters Marine Corps, 2011.

———. "Facebook Page." http://www.facebook.com/#!/marinecorps.

———. "Global Impact." http://www.marines.com/global-impact.

———. "Language, Regional and Culture Strategy 2011–2015." Headquarters Marine Corps, 2011.

———. "Lessons from 26th Marine Expeditionary Unit Operations." *Marine Corps Center for Lessons Learned* 7, no. 11 (November 2011).

———. "The Long War: Send in the Marines. A Marine Corps Operational Employment Concept to Meet an Uncertain Security Environment." Quantico, Va.: U.S. Marine Corps, 2008.

———. "Marine Corps Ranks." *Marines: The Official Site of the Marine Corps.* http://www.marines.mil/Marines/Ranks.aspx.

———. "Marine Corps Regional, Culture, and Language Familiarization Program." August 2011.

———. "Roles in the Corps: Infantry." www.marines.com/being-a-marine/roles-in-the-corps/ground-combat-element/infantry.

———. *Small Wars Manual*, FMFRP 12–15. Washington, D.C.: U.S. Government Printing Office, 1940.

———. "Vision and Strategy 2025." Foreward by General James T. Conway. 2008.

———. "Your Impact." http://www.marines.com/global-impact/your-impact.

U.S. Marine Corps Center for Advanced Operational Culture Learning (CAOCL). "2010 Survey of Marine Corps Attitudes to Culture and Language Training: Sample and Methods." CAOCL, 2010.

U.S. Marine Corps Combat Training Command. "Marine Training Film."

U.S. Marine Corps Community Services. "The Marine Corps: A Young and Vigorous Force, Demographic Update." 2012.

U.S. Marine Corps Force Structure Review Group. "Reshaping America's Expeditionary Force in Readiness: Report of the 2010 Marine Corps Force Structure Review Group." Quantico, Va.: U.S. Marine Corps, March 14, 2011.

U.S. Marine Corps Intelligence Activity. "C-GIRH Cultural Generic Information Requirements Handbook." August 2008.

U.S. Marine Corps MAGTF Staff Training Program (MSTP). *Red Cell–Green Cell*. MSTP Pamphlet 2–0.1 (October 2011).

U.S. Marine Corps Recruit Depot (MCRD). *Beginning the Transformation*. 2008.

———. "Modified 12 Week Recruit Training Schedule." August 2011.

U.S. Marine Corps Recruit Depot (MCRD) Parris Island. "Command Brief." 2011.

Vergano, Dan, and Elizabeth Weise. "Should Anthropologists Work Alongside Soldiers? Arguments Pro, Con Go to Core Values of Cultural Science." *USA Today*, December 9, 2008, 5D.

Walsh, Max. "New USMC Recruiting Video." http://www.youtube.com/watch?v=UFeHoMhuz7A.

Wedel, Janine R., Chris Shore, Gregory Feldman, and Stacy Lathrop. "Toward an Anthropology of Public Policy." *Annals of the American Academy of Political and Social Science* 600 (2005): 30–51.

Whitehead, R. Brian, and Neil L. Ferguson, eds. *War in the Tribal Zone: Expanding States and Indigenous Warfare*. Santa Fe, N.Mex.: School of American Research Press, 1992.

Whitelaw, Gavin. "Learning from Small Change: Clerkship and the Labors of Convenience." *Anthropology of Work Review* 29, no. 3 (2008): 62–69.

Woulfe, James B. *Into the Crucible: Making Marines for the 21st Century*. Novato, Calif.: Presidio Press, 1998.

Wright, Susan, ed. *Anthropology of Organizations*. London: Routledge, 1994.

Yanow, Dvorah. *How Does a Policy Mean? Interpreting Policy and Organizational Actions*. Washington, D.C.: Georgetown University Press, 1997.

Zimbardo, Philip. *The Lucifer Effect: Understanding How Good People Turn Evil*. New York: Random House, 2008.

Index